Ben Stacy Jerrik (Ed.)

Colwich, Staffordshire

Ben Stacy Jerrik (Ed.)

Colwich, Staffordshire

Great Haywood, Little Haywood, National Trust for Places of Historic Interest or Natural Beauty

Part Press

Imprint

Publisher:
Part Press is a trademark of
International Book Market Service Ltd., 17 Rue Meldrum, Beau Bassin, 1713-01 Mauritius
Email: info@bookmarketservice.com
Website: www.bookmarketservice.com

Published in 2012

Printed in: U.S.A., U.K., Germany. This book was not produced in Mauritius.

ISBN: 978-613-6-16174-7

Contents

Articles

References

Colwich,_Staffordshire

Colwich	
Population	**Expression error: "4,584" must be numeric**Template:Infobox UK place/trap[1] (2001 census)
District	Stafford
Shire county	Staffordshire
Region	West Midlands
Country	England
Sovereign state	United Kingdom
Post town	Stafford
Postcode district	ST
Police	Staffordshire
Fire	Staffordshire
Ambulance	West Midlands
EU Parliament	West Midlands
UK Parliament	Stafford

Colwich (*Old English 'Charcoal specialised-farm' or perhaps, 'Cola's specialised farm'*[2]) is a civil parish and village in Staffordshire, England. It is situated off the A51 road, about 3 miles (5 km) north west of Rugeley, and 7 miles (11 km) south east of Stafford. It lies principally on the north east bank of the River Trent near Wolseley Bridge, just north of Cannock Chase.

The parish comprises about 8800 acres (**unknown operator: u'strong'** km^2) of land in the villages and hamlets of Colwich, Great Haywood, Little Haywood, Moreton, Bishton and Wolseley Bridge.

Shugborough Hall was the ancestral home of the earls of Lichfield, four miles (6 km) NW by W of Rugeley. The estate was purchased by William Anson in the early 17th century and is now in the care of the National Trust. The Lichfield family vault is at at St Michael and All Angels Church in Colwich, a short distance from Shugborough Hall.[3]

Wolseley Hall was until recently the home of the Wolseley Baronets for at least eight centuries.

Colwich Abbey

The village is noted for Colwich Abbey of St Mary's, a community of Roman Catholic nuns of the English Benedictine Congregation founded in 1623 at Cambrai in the Spanish Netherlands. In 1836 the community, having been expelled from France during the French Revolution, finally settled at The Mount, Colwich, where they established the present house, raised to the rank of an abbey in 1928.

References

[1] Colwich Parish entry at Office of National Statistics, UK (http://www.neighbourhood.statistics.gov.uk/dissemination/LeadTableView.do?a=3&b=799511&c=colwich&d=16&e=15&g=486401&i=1001x1003x1004&m=0&r=1&s=1230907843418&enc=1&dsFamilyId=779)

[2] English Place Name Society Database at Nottingham University (http://www.nottingham.ac.uk/~cczappdv/epnnewmap/detailpop.php?placeno=9769)

[3] (http://news.bbc.co.uk/1/hi/uk/4450762.stm)

Great_Haywood

Great Haywood (52°48′N 2°00′W) is a village in central Staffordshire, England, just off the A51 about four miles from Rugeley.

Haywood Junction, Great Haywood

Great Haywood lies on the River Trent, where the Trent is met by its tributary, the River Sow. The village is also the site of a significant junction of the English inland canal network, Haywood Junction, where the Staffordshire and Worcestershire Canal meets the Trent and Mersey Canal. The waters around the village are widely regarded by guidebooks as some of the most attractive on the network.[1]

There are two churches, each of which has an attached school. St. John's RC School was classed as 'Good' in their most recent Ofsted inspection, and Anson CE School was deemed to be 'Outstanding' in December 2011. [2] [3]. Ofsted report on Anson Primary School, December 2011]</ref> [4]. St. Stephen's was designed by Thomas Trubshaw, and became the centre of a parish in 1858. St. John the Baptist's Catholic church was originally built in Tixall, about three miles away, as a private chapel to Tixall Hall, which was owned by the Aston family. When the estate was sold to Earl Talbot, the church was dismantled and rebuilt in Great Haywood.[5] The marks made on the blocks to allow reassembly can still be seen inside the church.

There was originally a mill and a brewery in the village, but both have been closed down and demolished, commemorated by the names of the roads where they once stood (Mill Lane and Brewery Lane). Following a fatal automobile accident in 1905, the mill pond was drained and the road straightened.

Great Haywood was served by a railway station which was opened by the North Staffordshire Railway on June 6, 1887 and closed in 1957.

In August 2002 advertisements were placed in the national press for an "hermit" to take up residence on the Great Haywood Cliffs above the nearby Shugborough estate, ancestral home of Lord Lichfield.[6] Fifty-five people applied, and Ansuman Biswas was chosen as hermit. Shugborough also serves as the headquarters of Staffordshire county's arts management team.

Great Haywood is the site of Essex Bridge, one of the largest surviving packhorse bridges in the country which stands over the river Trent near Shugborough Hall. It borders Cannock Chase, designated an area of outstanding natural beauty since 1958.

References

[1] Nicholsons Guide to the Waterways - Four Counties and Welsh Canals ISBN 978-0007211128
[2] http://www.ofsted.gov.uk/provider/files/1883719/urn/124379.pdf Ofsted report on St John's Catholic Primary, Nov. 2008]
[3] http://www.ofsted.gov.uk/provider/files/902517/urn/124353.pdf
[4] http://www.ofsted.gov.uk/provider/files/1883719/urn/124379.pdf
[5] 'History, Gazetteer and Directory of Staffordshire'
[6] The Guardian, August 2002 (http://www.guardian.co.uk/uk/2002/aug/20/arts.artsnews)

Bibliography

- White, William (1851). *History, Gazetteer and Directory of Staffordshire*. Sheffield.
- *Great War*: Garth, John (2003). *Tolkien and the Great War*. London: HarperCollins. ISBN 0-00-711952-6.

Little_Haywood

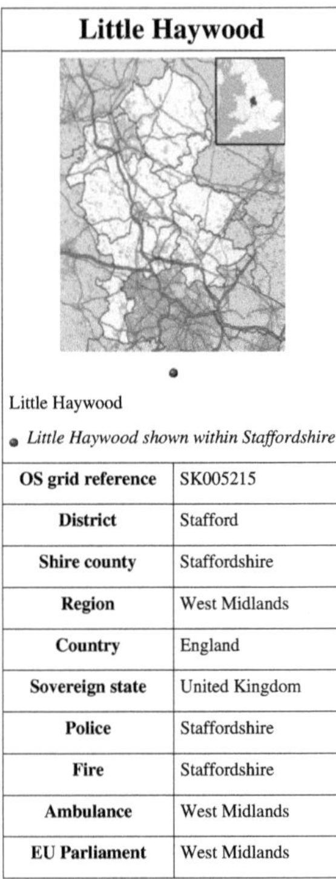

Little Haywood

Little Haywood

• *Little Haywood shown within Staffordshire*

OS grid reference	SK005215
District	Stafford
Shire county	Staffordshire
Region	West Midlands
Country	England
Sovereign state	United Kingdom
Police	Staffordshire
Fire	Staffordshire
Ambulance	West Midlands
EU Parliament	West Midlands

Little Haywood is a village in Staffordshire, England. It lies beside a main arterial highway, the A51 (linking the English Midlands with Liverpool) but traffic through the village is mainly light, owing to this bypass.[1] Nearby also is the West Coast Main Line railway, the Trent and Mersey Canal and beside it, the river Trent. Little Haywood is about 125 miles (**unknown operator: u'strong'** km) northwest of London, about 25 miles (**unknown operator: u'strong'** km) north of Birmingham, 4 miles (**unknown operator: u'strong'** km) northwest of Rugeley and 6 miles (**unknown operator: u'strong'** km) east of Stafford.

Little Haywood is cited in the Domesday Book of 1086.[2] Although originally a small village, housing expansion during the 1980s has created a commuter village and most of its inhabitants have employment far outside the confines of the Haywood area.

Location

Little Haywood is situated on the side of a hill in the system of valleys drained by the rivers Trent and Sow. It lies near the northern edge of Cannock Chase and is surrounded in the main by farmland. Geologically, the village lies on Triassic sandstone of the Sherwood Sandstone Group, with overlying glacial deposits from the last glaciation of Great Britain.

The village lies within the Stafford borough of Staffordshire and its name is derived from the Old English "haeg wadu," meaning an enclosure in woodland.

Waterways

There are three main waterways running near to Little Haywood: the River Sow, the River Trent and the Trent and Mersey Canal, which was opened in 1777. A wooden footbridge carrying Meadow Lane across the Trent was built in 1830. Previously the river was crossed by a ford, which was still used by cattle and horse-drawn vehicles after the footbridge was constructed. This wooden bridge was replaced by a brick- and stone-built Weetman's Bridge in 1887.[3] Less than 2 miles (**unknown operator: u'strong'** km) northwest of Little Haywood, the northeastern end of the Staffordshire and Worcestershire Canal joins the Trent and Mersey Canal. The Trent and Mersey Canal mile post at Little Haywood is number 37.[4]

Features and facilities

The Abbey Church of Saint Mary

The most prominent building in Little Haywood is St. Mary's Abbey. This Roman Catholic abbey is home to a community of enclosed Benedictine nuns and although part of the neighbouring Colwich parish, the abbey itself and its grounds lie alongside the road that runs through Little Haywood.

The Abbey Church of Saint Mary used to cover a large amount of Little Haywood and it has been said that there are underground tunnels leading from the abbey to Lichfield Cathedral, 10 miles (**unknown operator: u'strong'** km) away,[5] and to Shugborough Hall, a little over 1 mile (**unknown operator: u'strong'** km) away in the opposite direction.[6] Within the village, on land owned by Shugborough Hall, there is evidence of small-scale stone quarrying in the area known to locals as "the cliffs".[7]

Pubs and shops

The village and its outlying neighbours have an active parish community; the parish council organises events such as village fetes and on a day-to-day basis the social life of the village revolves around its public houses: the 'Red Lion' and the 'Lamb and Flag'.[8] There is no church in Little Haywood, no village green and no school. There are, however, the two pubs and a general store. The nearby village of Colwich is less than 1 mile (**unknown operator: u'strong'** km) away and has a church and a primary school but no pub or general store, and so amenities are often shared.

Wall

At the side of the road that runs from Little Haywood towards the village of Great Haywood, 1 mile (**unknown operator: u'strong'** km) away, is an example of a "make work wall", built by employees of Earl Talbot at Haywood Manor (no longer standing) during times when there was little else to do. In order to keep the workers from being idle, Talbot would make work for them in the form of features whose purpose might best be described as decorative.

J. R. R. Tolkien

The village was home to the newly-married Edith Tolkien, wife of author J. R. R. Tolkien, from March 1916 to February 1917.[9] Tolkien stayed with his wife in Cottage 1, Gipsy Green, on the Teddesley Park Estate, near the village during the winter of 1916, whilst recuperating from trench fever.[10] The surrounding landscape was said to be an inspiration for his early literary works about Middle-earth.. At the cottage he began work on what would become *The Silmarillion*. The village of Norbury lies about 14 miles (**unknown operator: u'strong'** km) away and may relate to the "Norbury of the Kings" that appears in *The Lord of the Rings*.

Tragic accidents at Colwich Junction

On September 18, 1986, two passenger trains collided at Colwich Junction, less than half a mile from Little Haywood, killing the driver of one of the trains and injuring 75 passengers. Several carriages of the crowded InterCity services were derailed.[11] On 2 January 2009, a Piper Cherokee single engine light aircraft came down at Colwich junction, killing the three people on board.[12]

References

[1] http://www.sabre-roads.org.uk/roadlists/f99/51.shtml
[2] http://www.staffshistory.org.uk/domesday.htm
[3] http://www.search.staffspasttrack.org.uk/engine/resource/default.asp?theme=422&originator=%2Fengine%2Ftheme%2Fdefault%2Easp&page=&records=&direction=&pointer=33&text=0&resource=13950
[4] tmc-mileposts.co.uk (http://www.tmc-mileposts.co.uk/MP_pages/milepost_37.html)
[5] http://www.enjoyengland.com/campaign/outdoor/aonbs/oe2_aonbs_cannock.aspx?bbcam=adwds_out&bbkid=Cannock+Chase&x
[6] http://www.shugborough.org.uk/
[7] http://atschool.eduweb.co.uk/anson.staffs/page15.html
[8] http://www.pubsulike.co.uk/newps/Staffordshire/Stafford.asp?Locality=Little+Haywood
[9] (*Great War* 2003, pg 134 & 231)
[10] (*Great War* 2003, pg 207)
[11] http://news.bbc.co.uk/onthisday/hi/dates/stories/september/19/newsid_2524000/2524593.stm
[12] http://news.bbc.co.uk/1/hi/england/staffordshire/7808254.stm

External links

- A map of Little Haywood dating from 1887 (http://www.old-maps.co.uk/oldmaps/index_external.jsp?easting=400500&northing=321500)
- The Staffordshire History Home Page (http://www.staffshistory.org.uk/)

National_Trust_for_Places_of_Historic_Interest_or_Nat

National Trust for Places of Historic Interest or Natural Beauty	
National Trust Logo	
Abbreviation	National Trust
Motto	For ever, for everyone
Formation	1894
Legal status	Trust
Purpose/focus	*To Look after Places of Historic Interest or Natural Beauty permanently for the benefit of the nation across England, Wales and Northern Ireland*
Headquarters	Swindon, United Kingdom
Location	United Kingdom
Official languages	English
Key people	H.R.H. The Prince of Wales (President) Dame Fiona Reynolds (Director-General) Sir Simon Jenkins (Chairman) Sir Laurie Magnus (Deputy Chairman)
Main organ	Board of Trustees
Affiliations	Various Organisations in the Council
Staff	4,964
Volunteers	61,000
Website	www.nationaltrust.org.uk [1]

The **National Trust for Places of Historic Interest or Natural Beauty**, usually known as the **National Trust**, is a conservation organisation in England, Wales and Northern Ireland. The Trust does not operate in Scotland, where there is an independent National Trust for Scotland.

According to its website:

> *"The National Trust works to preserve and protect the coastline, countryside and buildings of England, Wales and Northern Ireland.*
>
> *We do this in a range of ways, through practical caring and conservation, through educating and informing, and through encouraging millions of people to enjoy their national heritage."*[2]

The trust owns many heritage properties, including historic houses and gardens, industrial monuments and social history sites. It is one of the largest landowners in the United Kingdom, owning many beauty spots, most of which are open to the public free of charge. It is the largest membership organisation in the United Kingdom, and one of the largest UK charities by both income and assets.

History

The National Trust for Places of Historic Interest or Natural Beauty was incorporated in 1894 as an "association not for profit" under the Companies Acts 1862 to 1890, in which the liability of its members was limited by guarantee; it was later incorporated by six separate Acts of Parliament (The National Trust Acts 1907-1971 – as varied by a parliamentary scheme implemented by The Charities (National Trust) Order 2005),[3] it is also a charitable organisation registered under the Charities Act 1993.[4]

Its formal purpose is:

> The preservation for the benefit of the Nation of lands and tenements (including buildings) of beauty or historic interest and, as regards lands, for the preservation of their natural aspect, features and animal and plant life. Also the preservation of furniture, pictures and chattels of any description having national and historic or artistic interest.

The Trust was founded on 12 January 1894 by:

- Octavia Hill (1838–1912),
- Sir Robert Hunter (1844–1913) and
- The Very Rev. Hardwicke Canon Rawnsley (1851–1920),

prompted in part by the earlier success of Charles Eliot and the Kyrle Society. A fourth individual, The 1st Duke of Westminster (1825–1899), is also referred to in many texts as being a principal contributor to the formation of the Trust.

In the early days the Trust was concerned primarily with protecting open spaces and a variety of threatened buildings; its first property was Alfriston Clergy House and its first nature reserve was Wicken Fen. Its first archaeological monument was White Barrow.

The Trust's symbol, a sprig of oak leaves and acorns, is thought to have been inspired by a carving in the cornice of the Alfriston Clergy House.

Carving of an oak leaf at Alfriston Clergy House

The Trust has been the beneficiary of numerous donations of both property and money. However, probably the most bizarre were those given by a mysterious masked group known as Ferguson's Gang between about 1932 and 1940.

The focus on country houses and gardens which now comprise the majority of its most visited properties came about in the mid 20th century when it was realised that the private owners of many of these properties were no longer able to afford to maintain them. Many were donated to the Trust in lieu of death duties. The diarist James Lees-Milne is usually credited with playing a central role in the main phase of the Trust's country house acquisition programme, though he was in fact simply an employee of the Trust, and was carrying through policies which had already been decided by its governing body.

One of the biggest crises in the Trust's history erupted at the 1967 annual general meeting, when the leadership of the Trust was accused of being out of touch and placing too much emphasis on conserving country houses. In response, the Council asked Sir Henry Benson to chair an advisory Committee to review the structure of the trust. Following the publication of the Benson Report in 1968 much of the administration of the Trust was devolved to the regions.

Wicken Fen acquired by the National Trust in 1899

In 2005 the Trust moved to a new head office in Swindon, Wiltshire. The building was constructed on an abandoned railway yard, and is intended as a model of brownfield renewal. It is named Heelis, which is the married name of writer Beatrix Potter, who was one of the National Trust's most important benefactors.[5]

Governance

The Trust is an independent charity rather than a government institution (English Heritage and its equivalents in other parts of the United Kingdom are government bodies which perform some functions which overlap with the work of the National Trust).

It was founded as a not-for-profit company in 1895 but was later re-incorporated by a private Act of Parliament, the National Trust Act 1907. Subsequent Acts of Parliament between 1919 and 1978 amended and extended the Trust's powers and remit. In 2005 the governance of the Trust was substantially changed under a scheme made by the Charity Commission.[6]

The Trust is governed by a twelve-strong Board of Trustees. The Board is appointed and overseen by a Council which comprises 26 people elected by the members of the Trust, and 26 people appointed by other organisations whose work is related to that of the Trust, such as The Soil Association, the Royal Horticultural Society, and the Council for British Archaeology.[7]

At an operational level the Trust is organised into regions which are aligned with the official local government regions. Its headquarters are in Swindon. The Central Office building is Heelis, taken from the married name of children's author Beatrix Potter, a huge supporter and donor to the Trust.

Funding

For the year ended 28 February 2010, the Trust's total income was £406 million. The largest sources of income were membership subscriptions (£125.2 million), direct property income (£105.6 million) and legacies (£50.3 million). In addition, the Trust's commercial arm, National Trust Enterprises Ltd, which undertakes profit-making activities such as running gift shops and restaurants at properties, contributed £54.7 million.

Expenses included £177.4 million for routine property running costs and £100 million for capital projects.

The Trust is heavily supported by volunteers, who numbered about 61,000 in 2009/10, contributing 3.5 million hours of work with an estimated value of £29.2 million.

Membership

The Trust is one of the largest membership organisations in the world and annual subscriptions are its most important source of income. Membership numbers have grown from 226,200 when the Trust celebrated its 75th anniversary in 1970 to 500,000 in 1975, one million in 1981, two million in 1990 and by 2010, membership had reached 3.7 million. Members are entitled to free entry to trust properties that are open to the public at a charge.

The members elect half of the Council of the National Trust, and periodically (most recently in 2006) vote on the organisations which may appoint the other half of the Council. Members may also propose and vote on motions at the annual general meeting, although these are advisory and do not decide the policy of the Trust.

In the 1990s a dispute over whether stag hunting should be permitted on National Trust land caused bitter disputes within the organisation and was the subject of much debate at annual general meetings, but it did little to slow down the growth in member numbers.

There is a separate organisation called The Royal Oak Foundation for American supporters.

Volunteering

The National Trust was founded in 1895 by 3 volunteers. Last year the National Trust was helped by 61,000 volunteers. Volunteering experiences at the National Trust are varied, ranging from helping in historic houses and gardens to fundraising and providing specialist skills.[8] Thanks to the volunteering opportunities and schemes which the Trust runs for children, it is a member of The National Council for Voluntary Youth Services (NCVYS)[9] in recognition of its work for the personal and social development of young people.

National Trust properties

Historic houses and gardens

The Trust owns two hundred historic houses that are open to the public. The majority of them are country houses and most of the others are associated with famous individuals. The majority of these country houses contain collections of pictures, furniture, books, metalwork, ceramics and textiles that have remained in their historic context. Most of the houses also have important gardens attached to them, and the Trust also owns some important gardens not attached to a house. The properties include some of the most famous stately homes in the country and some of the key gardens in the history of British gardening.

A National Trust property sign at Gordale

The trust acquired the majority of its country houses in the mid 20th century, when death duties were at their most punitive. James Lees-Milne was secretary of the trust's Country House Committee in the key period either side of World War II. The arrangements made with families bequeathing their homes to the trust often allowed them to continue to live in part of the property. Since the 1980s the trust has been reluctant to take over large houses without substantial accompanying endowment funds, and its acquisitions in this category have been less frequent.

Coast and countryside

A National Trust signpost at Mill Dale on the River Dove at the start of the path to Tissington

The Trust's land holdings account for more than 630,000 acres (985 square miles, 2550 km²), mostly of countryside, covering nearly 1.5% of the total land mass of England, Wales and Northern Ireland. A large proportion of this consists of the parks and agricultural estates attached to country houses, but there are also many countryside properties which were acquired specifically for their scenic or scientific value. The Trust owns or has covenant over about a quarter of the Lake District; it has similar control over about 12% of the Peak District National Park (See for example South Peak Estate, High Peak Estate). It owns or protects roughly one fifth of the coast in England, Wales and Northern Ireland (704 miles, 1126 km), and has a long-term campaign, Project Neptune, which seeks to acquire more.

Cliffs and Worm's Head at Rhossili

Other properties

In recent years the Trust has sought to broaden its activities and appeal by acquiring properties such as former mills (early factories), workhouses and Paul McCartney and John Lennon's childhood homes.

Protection of Trust property

The National Trust Acts grant the Trust the unique statutory power to declare land *inalienable*. This prevents the land from being sold or mortgaged against the Trust's wishes without special parliamentary procedure. The inalienability of Trust land was overridden by Parliament in the case of proposals to construct a section of the Plympton by-pass through the park at Saltram, on the grounds that the road proposal had been known about before the park at Saltram was declared inalienable.[10]

The Acts also give the Trust the power to make bylaws to regulate the activities of people when on its land. All photography at National Trust properties, other than that for private and personal use or for entry into approved competitions, is strictly prohibited.[11] Since "private and personal use" means that they may not be displayed on the Internet,[12] visitors are instead directed to request images from the **National Trust Photo Library**.[13]

Most visited properties

2009–10

The 2009–10 annual report contains a list of all National Trust properties for which an admission charge is made that attracted more than 50,000 visitors in the year. The top ten were:

1. Wakehurst Place Garden — 439,627 (administered and maintained by the Royal Botanic Gardens, Kew)
2. Stourhead — 351,358
3. Waddesdon Manor — 348,308
4. Fountains Abbey & Studley Royal — 339,326
5. Attingham Park - 257,340
6. Polesden Lacey — 256,493
7. Belton House — 249,785
8. Carrick-a-Rede Rope Bridge — 248,609
9. Calke Abbey - 244,767
10. St Michael's Mount — 240,557

Stourhead's landscape garden.

Waddesdon Manor in Buckinghamshire

National Trust Places in the United Kingdom

- National Trust Properties in England
- National Trust Properties in Wales
- National Trust Properties in Northern Ireland

See also

- National Trust for Scotland
  - List of National Trust for Scotland properties
- Conservatoire du littoral
- Conservatoria delle Coste della Sardegna
- National Trust Magazine

References

[1] http://nationaltrust.org.uk/
[2] National Trust: What we do (http://nationaltrust.org.uk/main/w-trust/w-thecharity/w-what_we_do.htm)
[3] The National Trust Acts (http://www.nationaltrust.org.uk/main/w-nt_acts_1907-1971.pdf)
[4] Charity number 205846 - registration details (http://www.charitycommission.gov.uk/registeredcharities/showcharity.asp?chyno=205846) - retrieved on 16 January 2008
[5] National Trust: New central office: Heelis (http://nationaltrust.org.uk/main/w-trust/w-thecharity/w-new_central_office/w-new_central_office-heelis.htm)
[6] The National Trust Acts 1907 - 1971 as varied by a Parliamentary Scheme implemented by The Charities (National Trust) Order 2005 (http://nationaltrust.org.uk/main/w-nt_acts_1907-1971.pdf) - retrieved on 16 January 2008.
[7] Members of the Council of the National Trust (http://nationaltrust.org.uk/main/w-trust/w-thecharity/w-how_we_are_run/w-how_we_are_run-council/w-gove-council_members.htm) retrieved on 16 January 2008.
[8] National Trust | Volunteering | What's in it for you? (http://nationaltrust.org.uk/main/w-trust/w-volunteering/w-aboutvolunteering.htm)
[9] Full list of NCVYS members (http://www.ncvys.org.uk/index.php?page=392)
[10] History of the Trust: 1967 - 1994 (http://nationaltrust.org.uk/main/w-trust/w-thecharity/w-history_trust/w-history_trust-1967_1994.htm) - retrieved on 16 January 2008.
[11] FAQs at nationaltrust.org.uk (http://www.nationaltrust.org.uk/main/w-global/w-contact_us/w-faqs/w-faqs-visitor.htm#photo)
[12] National Trust Photography Persecution (http://www.wildaboutbritain.co.uk/forums/photography-techniques/52876-national-trust-photography-persecution.html)

[13] National Trust Photo Library (http://www.ntpl.org.uk/index2.pgi)

Bibliography

- Fedden, Robin, Joekes, Rosemary, "The National Trust Guide to England, Wales, and Northern Ireland", Norton, 1973. ISBN 0-393-01876-8.

External links

- The Preservation of Places of Interest or Beauty, (1907), lecture by Sir Robert Hunter
- The National Trust Acts 1907-71 (http://www.nationaltrust.org.uk/main/w-nt_acts_1907-1971.pdf)
- The National Trust website (http://nationaltrust.org.uk)
- The National Trust Conservatory Collection website (http://www.nationaltrustconservatories.co.uk)
- Annual report for 2004–05 (http://nationaltrust.org.uk/main/annual_report_0405_web_final1.pdf), including financial data (PDF document)
- Annual report for 2005–06 (http://nationaltrust.org.uk/main/w-annual_report-2006.pdf), including financial data (PDF document)
- Intelligent Giving profile of The National Trust (http://www.intelligentgiving.com/charity/205846)
- The Royal Oak Foundation website (http://www.royal-oak.org)
- The Official National Trust Print Website (http://www.ntprints.com/?ref=wiki&ad=nt001)

Video clips

- National Trust Vimeo channel (http://vimeo.com/channels/nationaltrust)
- National Trust YouTube channel (http://www.youtube.com/user/nationaltrustcharity)

Earl_of_Lichfield

Ditchley House, the seat of the Lee Family

Earl of Lichfield is a title that has been created three times in British history. Lord Bernard Stewart, youngest son of Esmé Stewart, 1st Duke of Lennox, was to be created Earl of Lichfield by Charles I for his actions at the battles of Newbury and Naseby but died before the creation could be implemented. Charles Stewart, the son of Bernard's younger brother George, who had been killed at the Battle of Edgehill, was instead created Earl of Lichfield in December 1645, soon after the Battle of Rowton Heath. Charles's cousin, who held the titles of Duke of Richmond and Earl of Lennox through the first Duke of Lennox's eldest son James, died aged eleven in 1660 with Charles as his heir. He married Frances Teresa Stuart, the celebrated beauty and alleged former mistress of King Charles II. In disgrace with the king, Charles was sent into exile as ambassador to Denmark, where he drowned on December 12, 1672. All of the English and Scottish titles that had been bestowed upon the male heirs became extinct.

The second creation came in 1674 when Charles II created Sir Edward Lee, 5th Baronet, of Quarendon, **Baron Spelsbury**, **Viscount Quarendon** and **Earl of Lichfield**. He married Charlotte Fitzroy, the illegitimate daughter of the King and Barbara Villiers, in 1677. He was succeeded by his third but eldest surviving son, George Henry Lee, the second Earl. He constructed the stately home of Ditchley in Oxfordshire. On his death the titles were passed on to his son George Henry Lee, the third Earl. He represented Oxfordshire in the House of Commons and served as Captain of the Honourable Band of Gentlemen Pensioners from 1762 to 1772. He died childless and was succeeded by his uncle, the fourth Earl. He was also childless. On his death in 1776 all his titles became extinct. The **Lee Baronetcy**, of Quarendon in the County of Buckingham, had been created in the Baronetage of England in 1611 for Henry Lee. He was the cousin and heir of Henry Lee of Ditchley.

The third creation came in 1831 in favour of Thomas Anson, 2nd Viscount Anson. The Anson family is descended from George Anson, Member of Parliament for Lichfield from 1770 to 1789. Born George Adams, he was the son of Sambrooke Adams and his wife Janette Anson, sister of the famous naval commander George Anson, 1st Baron Anson. In 1773, on the death of his uncle Thomas Anson (brother of Lord Anson), he succeeded to the substantial estates accumulated by his uncle Lord Anson, including the Anson family seat of Shugborough Hall in Staffordshire. The same year he assumed by Royal license the surname of Anson in lieu of Adams. His eldest son Thomas Anson represented Lichfield in the House of Commons as a Whig from 1789 to 1806. The latter year he was raised to the Peerage of the United Kingdom as **Baron Soberton**, of Soberton in the County of Southampton, and **Viscount Anson**, of Shugborough and Orgreave in the County of Stafford.

Shugborough Hall, the seat of the Anson family

He was succeeded by his eldest son, the second Viscount. He was also a Whig politician and served as Master of the Buckhounds from 1830 to 1834 and as Postmaster General from 1835 to 1841. In 1831 he was created **Earl of Lichfield**, of Lichfield in the County of Stafford, in William IV's coronation honours. This title is also in the Peerage of the United Kingdom. On death the titles passed to his eldest son, the second Earl. He sat as Member of Parliament for Lichfield and served as Lord-Lieutenant of Staffordshire. The titles descended from father to son until the death of his grandson, the fourth Earl, in 1960. He was succeeded by his grandson, the fifth Earl, the only son of Lieutenant-Colonel Thomas William Arnold Anson, Viscount Anson (1913–1958), eldest son of the fourth Earl. Known professionally as Patrick Lichfield, he was a successful photographer. As of 2010 the titles are held by his only son, the sixth Earl, who succeeded in 2005.

Several other members of the Anson family have also gained distinction. Sir George Anson, younger brother of the first Viscount, was a General in the Army and represented Lichfield in the House of Commons. His son Talavera Vernon Anson (1809–1895) was an Admiral in the Royal Navy. William Anson (1772–1847), younger brother of the first Viscount, was created a Baronet in 1831 (see Anson Baronets for more information on this branch of the family). The Very Reverend Frederick Anson (1779–1867), younger brother of the first Viscount, was Dean of Chester. The Hon. George Anson, second son of the first Viscount, was a prominent soldier and politician. The Hon. Augustus Anson, third son of the first Earl, was a Member of Parliament and recipient of the Victoria Cross. The Right Reverend the Hon. Adelbert John Robert Anson (1840–1909), fourth and youngest son of the first Earl, was a clergyman and served as Bishop of Qu'Apelle in Canada. The Hon. Sir George Augustus Anson (1857–1947), second son of the second Earl, was a courtier and Lieutenant-Colonel in the Army.

The courtesy title of the eldest son and heir apparent of the Earl is *Viscount Anson*.

The family seat is Shugborough Hall, Staffordshire, which is about 15 miles from the city of Lichfield. The Lichfield family vault is at at St Michael and All Angels Church in Colwich, a short distance from Shugborough Hall.[1]

Earls of Lichfield, First Creation (1645)

- Charles Stewart, 3rd Duke of Richmond, 6th Duke of Lennox, 1st Earl of Lichfield (1639–1672)

Lee Baronets, of Quarendon (1611)

- Sir Henry Lee, 1st Baronet (d. 1631)
- Sir Francis Henry Lee, 2nd Baronet (1616–1639)
- Sir Henry Lee, 3rd Baronet (1637–1658)
- Sir Francis Henry Lee, 4th Baronet (1639–1667)
- Sir Edward Henry Lee, 5th Baronet (1663–1716) (created **Earl of Lichfield** in 1674)

Earls of Lichfield, Second Creation (1674)

- Edward Henry Lee, 1st Earl of Lichfield (1663–1716)
  - Charles Lee, Viscount Quarendon (1680–1680)
  - Edward Henry Lee, Viscount Quarendon (1681–1713)
- George Henry Lee, 2nd Earl of Lichfield (1690–1742)
- George Henry Lee, 3rd Earl of Lichfield (1718–1772)
- Robert Lee, 4th Earl of Lichfield (1706–1776)

Viscounts Anson (1806)

- Thomas Anson, 1st Viscount Anson (1767–1818)
- Thomas William Anson, 2nd Viscount Anson (1795–1854) (created **Earl of Lichfield** in 1831)

Earls of Lichfield, Third Creation (1831)

- Thomas William Anson, 1st Earl of Lichfield (1795–1854)
- Thomas George Anson, 2nd Earl of Lichfield (1825–1892)
- Thomas Francis Anson, 3rd Earl of Lichfield (1856–1918)
- Thomas Edward Anson, 4th Earl of Lichfield (1883–1960)
  - Thomas William Arnold Anson, Viscount Anson (1913–1958)
- (Thomas) Patrick John Anson, 5th Earl of Lichfield (1939–2005)
- Thomas William Robert Hugh Anson, 6th Earl of Lichfield (b. 1978)

The heir apparent is the present holder's son, Thomas Ossian Patrick Wolfe Anson, Viscount Anson (b. 2011).[2]

See also

- Duke of Richmond
- Duke of Lennox
- Baron Anson
- Anson Baronets

References

[1] BBC NEWS | UK | Lichfield funeral date announced (http://news.bbc.co.uk/1/hi/uk/4450762.stm)
[2] (http://announcements.telegraph.co.uk/births/133768/lichfield)

- Kidd, Charles, Williamson, David (editors). *Debrett's Peerage and Baronetage* (1990 edition). New York: St Martin's Press, 1990.
- Leigh Rayment's Peerage Pages (http://www.leighrayment.com/)
- www.thepeerage.com (http://www.thepeerage.com/)
- Peerage of England, Arthur Collins, 1812 (http://books.google.com/books?id=yFI5AAAAMAAJ&pg=PP9&lpg=PP9&dq=anson+courtenay&source=web&ots=nSsXejMJuM&sig=G5xxH-MFWPdOjPHDq64ZjOghe5o&hl=en#PPA429,M1)

Wolseley_Baronets

There have been two **Baronetcies** created for members of the **Wolseley** family, one in the Baronetage of England and one in the Baronetage of Ireland. Both titles are extant as of 2008.

The **Wolseley Baronetcy**, of Wolseley in the County of Stafford, was created in the Baronetage of England on 24 November 1628 for Robert Wolseley, the member of an ancient Staffordshire family and a Colonel in Charles I's army. The second Baronet represented Oxfordshire, Staffordshire and Stafford in the House of Commons and was a member of Oliver Cromwell's House of Lords. The sixth Baronet was a Gentleman of the Privy Chamber to King George III.

The family seat was at Wolseley Park, Rugeley, Staffordshire. The old house Wolseley Hall was demolished in 1954 and the commercial ventures of the 11th Baronet created financial difficulties which led to the enforced sale of the estate in 1996[1]

The **Wolseley Baronetcy**, of Mount Wolseley in the County of Carlow, was created in the Baronetage of Ireland on 19 January 1745 for Richard Wolseley, who sat as a member of the Irish House of Commons for Carlow. He was the younger brother of the fifth Baronet of the 1628 creation. Consequently, the holder of the baronetcy is also in remainder to the Wolseley Baronetcy of Wolseley. As of 13 June 2007 the presumed thirteenth and present Baronet has not successfully proven his succession and is therefore not on the Official Roll of the Baronetage, with the baronetcy considered dormant since 1991. For more information, follow this link. [2]

Wolseley Baronets, of Wolseley (1628)

- Sir Robert Wolseley, 1st Baronet (c. 1587-1646)
- Sir Charles Wolseley, 2nd Baronet (c. 1630-1714)
- Sir William Wolseley, 3rd Baronet (c. 1660-1728)
- Sir Henry Wolseley, 4th Baronet (d. 1730)
- Sir William Wolseley, 5th Baronet (d. 1779)
- Sir William Wolseley, 6th Baronet (1740-1817)
- Sir Charles Wolseley, 7th Baronet (1769-1846)
- Sir Charles Wolseley, 8th Baronet (1813-1854)
- Sir Charles Michael Wolseley, 9th Baronet (1846-1931)
- Sir Edric Charles Joseph Wolseley, 10th Baronet (1886-1954)
- Sir Charles Garnet Richard Mark Wolseley, 11th Baronet (b. 1944)

Wolseley, of Mount Wolseley (1745)

- Sir Richard Wolseley, 1st Baronet (1696-1769)
- Sir Richard Wolseley, 2nd Baronet (1729-1781)
- Sir William Wolseley, 3rd Baronet (1775-1819)
- Sir Richard Wolseley, 4th Baronet (1760-1852)
- Sir Clement Wolseley, 5th Baronet (1794-1857)
- Sir John Richard Wolseley, 6th Baronet (1834-1874)
- Sir Clement James Wolseley, 7th Baronet (1837-1889)
- Sir John Wolseley, 8th Baronet (1803-1890)
- Sir Capel Charles Wolseley, 9th Baronet (1870-1923)
- Sir Reginald Beatty Wolseley, 10th Baronet (1872-1933)
- Sir William Augustus Wolseley, 11th Baronet (1865-1950)
- Sir Garnet Wolseley, 12th Baronet (1915-1991)
- *Sir James Douglas Wolseley, 13th Baronet* (b. 1937)

Notes

[1] *The Daily Telegraph, Features p27* 2 April 2008
[2] http://www.baronetage.org/succession-to-baronetcy/

References

- Kidd, Charles, Williamson, David (editors). *Debrett's Peerage and Baronetage* (1990 edition). New York: St Martin's Press, 1990.
- Leigh Rayment's List of Baronets (http://www.leighrayment.com/baronetage.htm)

English_Benedictine_Congregation

The **English Benedictine Congregation** (abbr. EBC) comprises autonomous Roman Catholic Benedictine communities of monks and nuns and is technically the oldest of the 21 congregations that are affiliated in the Benedictine Confederation.

History and administration

Although the EBC claims technical canonical continuity with the congregation erected by the Holy See in 1216, that earlier English Congregation was destroyed at the dissolution of the monasteries in 1535-40. The present English Congregation was revived and restored by Rome in 1607-33 when numbers of Englishmen and Welshmen had become monks in continental European monasteries and were coming to England as missioners.

At the beginning of the 21st century the EBC has Houses in the United Kingdom, the United States, South America and Africa.

Every four years the General Chapter of the EBC elects an Abbot President from among the Ruling Abbots with jurisdiction, and those who have been Ruling Abbots. He is assisted by a number of officials. Periodically he undertakes a Visitation of the individual Houses. The purpose of the Visitation is the preservation, strengthening and renewal of the religious life, including the laws of the Church and the Constitutions of the congregation. The President may require by Acts of Visitation, that particular points in the Rule, the Constitutions and the law of the Church be observed.

The current Abbot President is the Right Reverend Dom Richard Yeo, former Abbot of Downside Abbey.

Houses of the English Benedictine Congregation

Houses of the Congregation in exile

Religious house in Europe	Location	Dates	Successor house in England
St. Gregory's Priory, Douai	Douai, France	1607–1798	Downside Abbey
Dieulouard Priory	France	1608–1798	Ampleforth Abbey
St. Malo Priory	St. Malo, Brittany	1610 approx. -late 17th century	n/a
St. Edmund's Priory, Paris; later St. Edmund's Abbey, Douai	Paris	1615-1798 (Paris); 1818-1903 (Douai)	Douai Abbey, Woolhampton
Cambrai Priory	Cambrai, Flanders	1625–1794	Stanbrook Abbey
Our Lady of Good Hope Priory, Paris	Paris	1651–1794	Colwich Abbey
Lamspringe Abbey	Lamspringe, Lower Saxony	1630–1803	Broadway Priory, 1826–34; Fort Augustus Abbey, 1886–1998

Houses of the present Congregation

United Kingdom:

- Ampleforth Abbey, fdd 1608 at Dieulouard
- Belmont Abbey, fdd 1859
- Buckfast Abbey, fdd 1882
- Colwich Abbey (nuns), fdd 1651 in Paris
- Curzon Park Abbey (nuns), fdd 1868
- Douai Abbey, fdd 1615 in Paris
- Downside Abbey, fdd 1607 in Douai
- Ealing Abbey, fdd 1897
- Stanbrook Abbey (nuns) fdd 1625 in Cambrai
- Worth Abbey, fdd 1933

United States:

- Portsmouth Abbey, fdd 1918
- Saint Louis Abbey, fdd 1955
- Saint Anselm's Abbey, fdd 1923

Zimbabwe

- Monastery of Christ the Word, fdd 1996

Peru

- Priory of the Incarnation

Notes

Sources

- Website of the EBC (http://www.benedictines.org.uk/)
- History of the EBC (http://www.benedictines.org.uk/history.htm)
- *Religiosus Ordo*, The Apostolic Letter of Pope Leo XIII of 12 November 1889, concerning the modification of the government and discipline of the EBC (http://www.catholic-history.org.uk/ebc/doc_religiosus.htm)(with an outline of the EBC history since the 16th century)
- *The Benedictine Yearbook 2005*

Colwich_Abbey

Colwich Abbey is a community of Roman Catholic nuns of the English Benedictine Congregation founded in 1623 at Cambrai, Flanders, in the Spanish Netherlands. During the French Revolution, the community was expelled from France, and settled at The Mount, Colwich, Staffordshire, in 1836, where it continues today.

History

St Mary's Abbey of English Benedictine nuns has its origins in 1623 at Cambrai in the Spanish Netherlands. At that time, persecution made it impossible for women to become nuns in England. By 1645, the Cambrai community under Abbess Catherine Gascoigne had increased to 50 nuns, and was living in conditions of extreme poverty.

On 6 February 1652, the community was established in Paris as the Priory of Our Lady of Good Hope under Dame Bridget More as their Prioress. She was a direct descendant of the martyr, St Thomas More, and had been taught at Cambrai under the spiritual supervision of the great mystical theologian, Dom Augustine Baker.

In the French Revolution, the abbey was suppressed and the nuns were imprisoned, first in the monastery and then in the Château de Vincennes. When released in 1795, they settled in England, first in Dorset and then at Cannington in Somerset. In 1836, they finally settled at The Mount, Colwich, where they named the new house "St. Benedict's Priory".

In 1928, the Priory was raised to the rank of an Abbey, and the house was re-named St. Mary's Abbey. A daughter house, at Atherstone in Warwickshire, continued as a separate community until 1967.

External links

- St Mary's Abbey, Colwich [1]

See also

- List of abbeys and priories

References

[1] http://www.colwichabbey.org.uk/

Shugborough_Hall

Shugborough is a country estate in Great Haywood, Staffordshire, England, 4 miles from Stafford on the edge of Cannock Chase. It comprises a country house, kitchen garden, and model farm. Owned by the National Trust and maintained by the leaseholder, Staffordshire County Council, it previously belonged to the Earls of Lichfield, the Anson family.

Shugborough Hall today

History

Shugborough Hall in the 1820s.

The Shugborough estate was owned by the Bishops of Lichfield until the Dissolution of the Monasteries and therefter passed through several hands until it was purchased in 1624 by William Anson, a lawyer, of Dunston, Staffordshire[1]

In about 1693 his grandson William Anson (1656–1720) demolished the old house and created a new mansion.[1] The entrance front then to the west, comprised a ballustraded three storey, seven bayed central block . In about 1748 his great grandson Thomas Anson commissioned architect Thomas Wright to remodel the house, which was extended with flanking two storey, three bayed pavilions linked to the central block by pedimented passages.[1] At the turn of the 18th century the house was further altered and extended by architect Samuel Wyatt, when the pavilions and passages were incorporated into the main building and a new porticoed entrance front with ten Doric order pillars was created at the east.[1] for Thomas Anson, the 1st Viscount Anson and his wife Anne Margaret Coke, daughter of Thomas Coke, the 1st Earl of Leicester, whom he married in 1794. Styled Viscountess Anson in 1806, Anne Margaret Coke Anson died in London in 1843 and was buried at Shugborough.

Around 1750 the architect James "Athenian" Stuart, created a number of follies and monuments in the grounds. These include the Tower of Winds (based on one in Greece), the Chinese House (a Chinese-style pagoda), a triumphal arch based on Hadrian's, a Doric Temple, the Cat's Monument, and the Shepherd's Monument.

The grounds are connected to the village of Great Haywood by the Essex Bridge, built in the Middle Ages, and contain numerous sculptures in addition to Stuart's follies.

Nearby is Milford Hall, the estate of the Levett Haszard family, who are related to the Ansons and who sit on the board at Shugborough.[2]

Family History

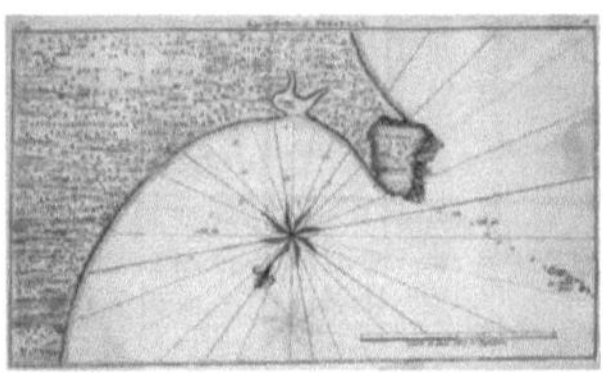

Illustration from French volume illustrating George Anson's voyage around the world

The Anson family who purchased the estate in the 17th century from Thomas Whitby of Great Haywood, Staffordshire produced some famous men, including Admiral George Anson, 1st Baron Anson, George Anson (British soldier), General George Anson (1769-1849), Thomas Anson (MP), Dean of Chester Frederick Anson and his sons George Edward Anson and Frederick Anson, Canon of St George's Chapel at Windsor Castle. Seven ships in the Royal Navy have been christened HMS *Anson*, honouring the first Baron Anson's circumnavigation in the 1740s.

The house contains a collection of photographs by the house's recent resident, the royal photographer, the late Patrick Anson, 5th Earl of Lichfield (1939-2005). Through his mother Anne (1917–1980), he was a first cousin, once removed, of Queen Elizabeth II of the United Kingdom, his mother having been a niece of Elizabeth Bowes-Lyon, the late Queen Mother. The 5th Earl of Lichfield married in 1975 Lady Leonora Grosvenor, daughter of the 5th Duke of Westminster. After divorcing in 1986, the Countess of Lichfield retained her title and has not remarried.

The Shepherd's Monument

The Shugborough inscription, still unsolved

The Shepherd's Monument has been internationally well-known since 1982, when the book *The Holy Blood and the Holy Grail* drew attention to the mysterious Shugborough inscription. Carved by an unknown 18th-century craftsman, this has been called one of the world's top uncracked ciphertexts.[3] [4] Theories have abounded, including some which suggest it may indicate the whereabouts of the Holy Grail, an idea fuelled by the Anson family's ancestral ties to the Knights Templar.

In January 2011 the British press revealed that A. J. Morton had solved the code. The letters O.U.O.S.V.A.V.V. & D.M., the Times explained, were probably created for, by, or in memorial of, Viscount Anson and his wife Mary Vernon-Venables.[5]

In recent years, codebreakers from the National Codes Center at Bletchley Park in Buckinghamshire have tried unsuccessfully to decipher it.[6] Before them, it is said that Charles Darwin and Charles Dickens also tried, and similarly failed.

Numerous explanations have been put forward, linking the code to the Priory of Sion, the Holy Grail and UFO's.

One more modest and romantic theory is that the inscription is a secret message between two lovers.[7]

Noted guests

Ornamental copy of Nicolas Poussin's Arcadia at Shugborough

J. R. R. Tolkien, author of *The Lord of the Rings*, stayed in Great Haywood during the winter of 1916/17 and in his story 'The Tale of the Sun and the Moon' (*The Book of Lost Tales 1*) he writes about a gnome called Gilfanon who owned an ancient house "...the House of a Hundred Chimneys, that stands nigh the bridge of Tavrobel". Tavrobel he describes as a village near the confluence of two rivers. If you stand on the Essex Bridge, you can see where the river Sow meets the river Trent, and Shugborough Hall has about 80 chimneys.

Another fantasy author, Mark Chadbourn, features Shugborough and the mysterious bas-relief on the Shepherd's Monument in his novel *The Hounds of Avalon,* part of The Dark Age sequence. In the novel, the gardens provide a point of access to the magical Otherworld of Celtic mythology.

Nicolas Poussin's Arcadia and the inscription also figure prominently in the fiction work by Steve Berry, *The Alexandria Link*. They are used to find the location of the Library of Alexandria.

The Present Day

The estate was gifted to the National Trust by the Anson family in 1960 in lieu of death duties[1] : it is managed on behalf of the owners by the Staffordshire County Council. The family resided in private apartments in the house until April 2010. Following the death of Patrick Lichfield on 11 November 2005 the private apartments were opened to the public in March 2011 where they can be viewed during a visit to the house.

The grounds and mansion house are open to the public. The attraction is marketed as "The Complete Working Historic Estate", which includes a working model farm museum dating from 1805 complete with a working watermill, kitchens, a dairy, a tea room, and rare breeds of farm animals. The walled garden, also dating from 1805, was restored in 2006 and also forms part of the attraction.

In addition, the house contains the historic servants' quarters and, within these, a Brewery. Originally restored in 1990, this is England's only log-fired brewery that still produces beer commercially, through a partnership with a local brewery. Previously used only on special occasions, the brewhouse has been a working exhibit since 2007.

See also

- List of Grade I listed buildings in Staffordshire

References

[1] *Shugborough* by Gervase Jackson-Stops for the National Trust (1980)

[2] The Genealogy of the Existing British Peerage and Baronetage: Containing the Family Histories of the Nobility, Edmund Lodge, Norroy King of Arms, London, 1859 (http://books.google.com/books?id=JC0BAAAAQAAJ&pg=PA331&lpg=PA331&dq=levett+shugborough+anson&source=web&ots=ZOLWygcdN2&sig=H6afRwg4juMZzPzHftYmjCEzRM8&hl=en)

[3] "Top 10 Uncracked Codes" (http://listverse.com/miscellaneous/top-10-uncracked-codes). The List Universe. . Retrieved 2008-11-24.

[4] Belfield, Richard (August 2007). *The Six Unsolved Ciphers: Inside the Mysterious Codes That Have Confounded the World's Greatest Cryptographers*. Ulysses Press. ISBN 1-56975-628-7.

[5] J. McNee, "Irvine Historian May Have Solved Ancient Puzzle" (http://www.irvinetimes.com/), Irvine Times Jan 26 2011 p.1 & 12

[6] New Puzzle for Code Breakers, BBC News, bbc.co.uk (http://news.bbc.co.uk/2/hi/uk_news/england/staffordshire/3703191.stm)

[7] Bell, David (2005). "12". *Staffordshire Tales of Murder & Mystery*. Murder & Mystery. Countryside Books. pp. 108. ISBN 1-85306-922-1.

External links

- Official website of Shugborough (http://www.shugborough.org.uk/)
- Memorial service for Patrick Anson (http://www.timesonline.co.uk/article/0,,61-2078356,00.html)
- Shugborough Garden - information on garden history and design (http://www.gardenvisit.com/g/shug.htm)
- Shugborough information at the National Trust (http://www.nationaltrust.org.uk/main/w-vh/w-visits/w-findaplace/w-shugboroughestate/)
- Stafford Tourism Bureau (http://www.visitstafford.org)
- The Return of The Christ? (http://andrewgough.co.uk/forum/viewforum.php?f=18)
- "Lord Lichfield's photographs to go on display at Shugborough Hall" (http://www.guardian.co.uk/artanddesign/2011/mar/18/lord-lichfield-photograph-exhibition), Maev Kennedy, *The Guardian*, Retrieved 03-18-2011

Moreton,_Staffordshire

Moreton is a small rural village in Staffordshire, England. It lies 3.5 miles (**unknown operator: u'strong'** km) south-west from Gnosall railway station, and 4 miles (**unknown operator: u'strong'** km) south-east from Newport, both on the Stafford and Shrewsbury section of the former London and North Western Railway.

Description

Two notable sites in the village are the village community hall built in 2000 and St. Mary's Church. The church of St. Mary is a stone building, in the Italian style, and was erected in 1837; it consists of chancel, transepts and nave, with tower and one bell, and seats 340 people. One other local site which is now a privately-owned home is the school, which at one point had over one hundred students from Moreton and its hamlets.

Considering the size of the village (roughly 50-60 houses), the lack of more facilities is understandable.

The village is made up primarily of three roads: Heath Lane, Post Office Lane and Church Lane. Both the village hall and church are located on Church Lane, nearer to where the Lane joins Post Office Lane.

History

Moreton ecclesiastical parish was formed on April 26, 1845, from the parish of Gnosall, although it remained in Gnosall parish for civil purposes.

Until the mid 1870s, it was referred to as a hamlet of Gnosall , and can be found, prior to 1880, described with the history of Gnosall. However, by 1880 it had developed sufficiently to be referred to separately, with its own 'satellite' hamlets: **Bromstead Heath**, Great Chatwell, **Coley**, Outwoods and **Wilbrighton**.

External links

- Map sources for Moreton, Staffordshire

A51_road

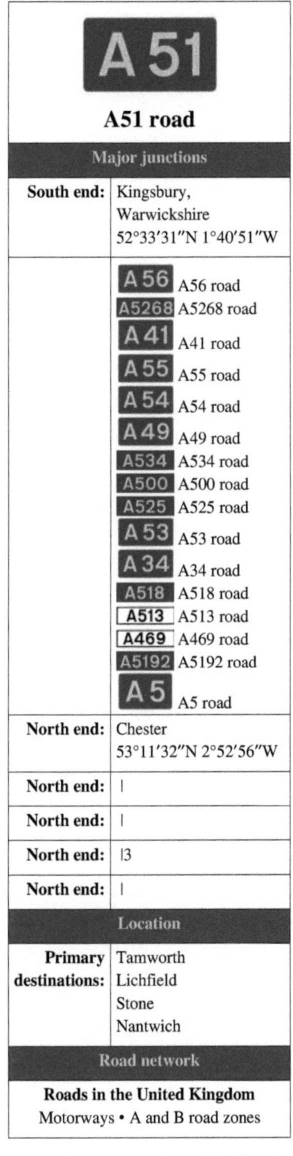

A51 road	
Major junctions	
South end:	Kingsbury, Warwickshire 52°33′31″N 1°40′51″W
	A56 A56 road A5268 A5268 road A41 A41 road A55 A55 road A54 A54 road A49 A49 road A534 A534 road A500 A500 road A525 A525 road A53 A53 road A34 A34 road A518 A518 road A513 A513 road A469 A469 road A5192 A5192 road A5 A5 road
North end:	Chester 53°11′32″N 2°52′56″W
North end:	I
North end:	I
North end:	I3
North end:	I
Location	
Primary destinations:	Tamworth Lichfield Stone Nantwich
Road network	
Roads in the United Kingdom Motorways • A and B road zones	

The **A51** is a road in England running from Kingsbury in Warwickshire (just east of Birmingham) to Chester. The road follows the following route:

- Kingsbury
- Tamworth

- Lichfield
- Rugeley (bypass opened 2007 [1])
- Little Haywood
- Great Haywood
- Weston
- Sandon
- Stone (merges briefly with A34)
- Woore
- Nantwich
- Tarporley (merges briefly with A49)
- Clotton
- Duddon
- Tarvin
- Littleton
- Vicars Cross
- Chester

The A51 is used by some long-distance traffic as an alternative to the M6 motorway, which is prone to congestion through Birmingham. The nearby M6 Toll motorway now serves a similar function.

References

[1] http://www.rugeleybypass.co.uk/

Southern_Netherlands

<table>
<tr><th colspan="4">History of the Low Countries</th></tr>
<tr><td colspan="3">Frankish Kingdom
(5th to 10th century)</td><td>Frisian Kingdom
(600-734)</td></tr>
<tr><td colspan="4">Carolingian Empire after 800</td></tr>
<tr><td></td><td>West Francia ("France")</td><td colspan="2">Independent Kingdom of Middle Francia (Lotharingia)
(843–870)</td></tr>
<tr><td colspan="4">Flanders and Lotharingia in Kingdom of West Francia
(870–880)</td></tr>
</table>

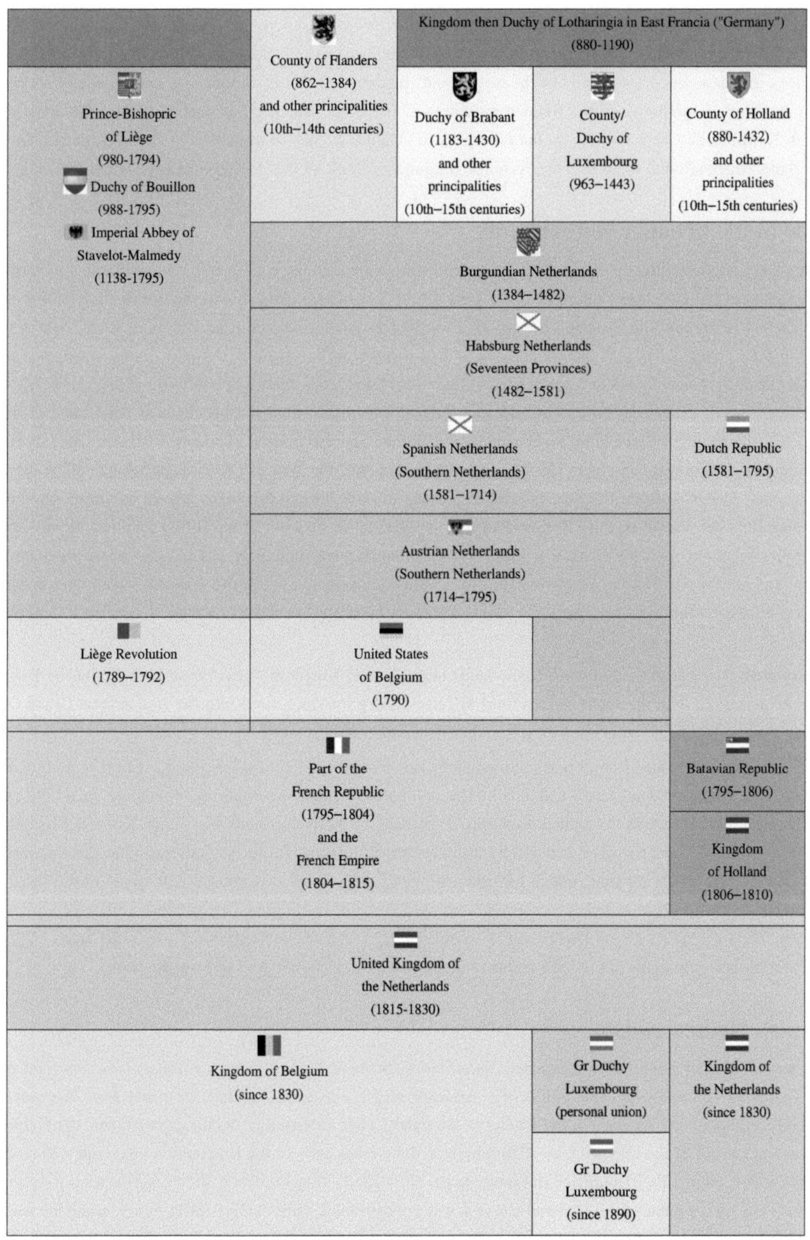

The **Southern Netherlands** (Dutch: *Zuidelijke Nederlanden*, Spanish: *Países Bajos del Sur*, French: *Pays-Bas méridionaux*) were part of the Low Countries controlled by Spain (**Spanish Netherlands**, Dutch: *Spaanse Nederlanden*, 1579–1714), Austria (**Austrian Netherlands**, Dutch: *Oostenrijkse Nederlanden*, German: *Österreichische Niederlande*, 1714–94) and annexed by France (1794–1815). This region comprised most of modern

Belgium (excepting three Lower-Rhenish territories: the Prince-Bishopric of Liège, the Imperial Abbey of Stavelot-Malmedy and the County of Bouillon) and Luxembourg (including the homonymous present Belgian province), and in addition some parts of the Netherlands (namely the Duchy of Limburg, now divided between the Dutch province of Limburg and the Belgian provinces of Liège and Limburg) as well as, until 1678, most of the present Nord-Pas-de-Calais region in northern France. Unlike French Burgundy and the republican Northern Netherlands, these allodial states kept access to the Burgundian Circle of the Holy Roman Empire until its end.

Place in the broader Netherlands

As they were very wealthy, the Netherlands in general were a jewel in the ever debt-burdened Habsburg crown, but unlike others of the Habsburg dominions, they were led by a merchant class. It was the merchant economy which made them wealthy and the Spanish attempts at increasing taxation, to finance the Habsburg wars[1], was a major factor in their proud defence of ancient privileges. This together with resistance to the religious intolerance of the Catholic Spanish monarchy led to a general rebellion of the Netherlands against Spanish rule in the 1570s. Although the northern seven provinces, led by Holland and Zeeland, established their independence as the United Provinces after 1581, the southern Netherlands were reconquered by the Spanish general Alexander Farnese, Duke of Parma. The Southern Netherlands passed to the Austrian Habsburgs after the War of the Spanish Succession in the early 18th century. Under Austrian rule, the provinces' defence of their ancient privileges proved as troublesome to the reforming Emperor Joseph II, Holy Roman Emperor as it had to his ancestor Philip II two centuries before, leading to a major rebellion in 1789–1790. The Austrian Netherlands were ultimately lost to the French Revolutionary armies, and annexed to France. Following the war, Austria's loss of the territories was confirmed, and they were joined with the northern Netherlands as a single kingdom under the House of Orange at the 1815 Congress of Vienna.

The Congress first joined the Southern Netherlands to the United Kingdom of the Netherlands under the House of Orange-Nassau, but with the south-eastern third of Luxembourg Province made into the autonomous Grand Duchy of Luxembourg, because it was claimed by both the Netherlands and Prussia.

In 1830 the predominantly Roman Catholic southern half became independent as the Kingdom of Belgium (the northern half being predominantly Calvinist). The autonomy of Luxembourg was recognised in 1839, but an instrument to that effect was not signed until 1867. The King of the Netherlands was Grand Duke of Luxembourg until 1890, when William III was succeeded by his daughter, Wilhelmina of the Netherlands – but Luxembourg still followed the Salic law at the time, which forbade a woman to rule in her own right, so the union of the Dutch and Luxembourger crowns then ended. The north-western two-thirds of the original Luxembourg remains a province of Belgium. The flags of the Grand Duchy of Luxembourg ( ) and of the Kingdom of the Netherlands ( ) are still distinguished only in the tint of their colours (although the former is not derived from the latter).

Spanish Netherlands

The **Spanish Netherlands** (Dutch: *Spaanse Nederlanden*, Spanish: *Países Bajos españoles*) was a portion of the Low Countries controlled by Spain from the sixteenth to the eighteenth century, inherited from the Dukes of Burgundy. Although the territory of the Duchy of Burgundy itself remained in the hands of France, the Habsburgs remained in control of the title of Duke of Burgundy and the other parts of the Burgundian inheritance, notably the Low Countries and the Free County of Burgundy in the Holy Roman Empire. They often used the term Burgundy to refer to it (e.g. in the name of the Imperial Circle it was grouped into), until the late 18th century, when the Austrian Netherlands were lost to the French Republic.

When part of the Netherlands separated from Spanish rule and became the United Provinces in 1581 the remainder of the area became known as the Spanish Netherlands and was still under the control of Spain. This region comprised modern Belgium, Luxembourg as well as part of northern France.

The Spanish Netherlands originally consisted of the whole of the

- county of Flanders, including Lilloise Flanders
- county of Artois
- city of Tournai
- Cambrai (roughly the département Nord and the northern half of Pas-de-Calais in modern France)
- duchy of Luxembourg
- duchy of Limburg
- county of Hainaut
- county of Namur
- heerlijkheid of Mechelen[2] (officially a county since 1490)
- duchy of Brabant, including the Margraviate of Antwerp
- the Upper Quarter (*Bovenkwartier*) of the duchy of Guelders (around Venlo and Roermond, in the present province of Dutch Limburg, and the town of Geldern in the present German district Kleve)

The capital was Brussels in Brabant.

In the early seventeenth century, there was a flourishing court at Brussels, which was under the government of King Philip III's half-sister Archduchess Isabella and her husband, Archduke Albert of Austria. Among the artists who emerged from the court of the "Archdukes", as they were known, was Peter Paul Rubens. Under the Archdukes, the Spanish Netherlands actually had formal independence from Spain, but always remained unofficially within the Spanish sphere of influence, and with Albert's death in 1621 they returned to formal Spanish control, although the childless Isabella remained on as Governor until her death in 1633.

The failing wars intended to regain the 'heretical' Northern Netherlands meant significant loss of (still mainly Catholic) territories in the north, which was consolidated in the 1648 Westphalian peace, and given the peculiar, inferior status of *Generality Lands* (jointly ruled by the United Republic, not admitted as member provinces) : Zeeuws-Vlaanderen (south of the river Scheldt), the present Dutch province of Noord-Brabant and Maastricht (in the present Dutch province of Limburg).

In the wars between the French and the Spanish in the seventeenth century, the territory of the Spanish Netherlands was repeatedly invaded, with portions seized by France. The French annexed Artois and Cambrai by the Treaty of the Pyrenees of 1659, and Dunkirk was ceded to the English. By the Treaties of Aix-la-Chapelle (ending the War of Devolution in 1668) and Nijmegen (ending the Franco-Dutch War in 1678), further territory up to the current Franco-Belgian border was ceded, including Lilloise Flanders (around the city of Lille), as well as half of the county of Hainaut (including Valenciennes). Later, in the War of the Reunions and the Nine Years' War, France annexed other parts of the region.

Austrian Netherlands

Under the Treaty of Utrecht (1713), following the War of the Spanish Succession, what was left of the Spanish Netherlands was ceded to Austria and thus became known as the **Austrian Netherlands**. However, the Austrians themselves generally had little interest in the region (aside from a short-lived attempt by Emperor Charles VI to compete with British and Dutch trade through the Ostend Company), and the fortresses along the border (the Barrier Fortresses) were, by treaty, garrisoned with Dutch troops. The area had, in fact, been given to Austria largely at British and Dutch insistence, as these powers feared potential French domination of the region.

A map of the dominion of the Habsburgs following the Battle of Mühlberg (1547) as depicted in *The Cambridge Modern History Atlas* (1912); Habsburg lands are shaded green. From 1556 the dynasty's lands in the Low Countries, the east of France, Italy, Sardinia, and Sicily were retained by the Spanish Habsburgs.

Throughout the latter part of the eighteenth century, the principal foreign policy goal of the Habsburg rulers was to exchange the Austrian Netherlands for Bavaria, which would round out Habsburg possessions in southern Germany. In the 1757 Treaty of Versailles, Austria agreed to the creation of an independent state in the Southern Netherlands ruled by Philip, Duke of Parma and garrisoned by French troops in exchange for French help in recovering Silesia. However the agreement was later revoked by the Third Treaty of Versailles and Austrian rule continued.

The Low Countries (with Liège, Stavelot-Malmedy and Bouillon) until 1795

In 1784 Joseph II, Holy Roman Emperor did take up the long-standing grudge of Antwerp, whose once-flourishing trade was destroyed by the permanent closing of the Scheldt, and demanded that the Dutch Republic open the river to navigation. However, the Emperor's stance was far from militant, and he called off hostilities after the so-called Kettle War, known by that name because its only "casualty" was a kettle. Though Joseph did secure in the 1785 Treaty of Fontainebleau that the Southern Netherlands would be compensated by the Dutch Republic for the continued closing the Scheldt, this achievement failed to gain him much popularity.

The Austrian Netherlands rebelled against Austria in 1788 as a result of Joseph II's centralizing policies. The different provinces established the United States of Belgium (January 1790). Austrian imperial power was restored by Joseph's brother and successor, Leopold II by the end of 1790.

French annexation

After the French Revolution, in 1794 the entire region (including territories that were never under Habsburg rule, like the Bishopric of Liège) was overrun by France ending the existence of this territory as Spanish/Austrian Netherlands. This was resisted by the Flamingant movement organized by Roman Catholic clergy. It became an integral part of France, and was divided into départements:

- Deux-Nèthes
- Dyle
- Escaut
- Forêts

- Jemmape
- Lys
- Meuse-Inférieure
- Ourthe
- Roer
- Sambre-et-Meuse

Austria confirmed the loss of its territories by the Treaty of Campo Formio, in 1797.

After the defeat of Napoleon in 1815 the region was given to the United Kingdom of the Netherlands, but after the Belgian Revolution of 1830 it separated to become the independent state of Belgium.

See also

- List of Governors of the Habsburg Netherlands
- List of plenipotentiaries of Austrian Netherlands
- Seventeen Provinces
- Union of Atrecht (Including map, 1579)
- Spanish Armada

Footnote

- Note 1: The example of these expensive wars which is best known to English-speaking people is that of the Spanish Armada. However, that came in 1588, a little after the Dutch had become exasperated to the extent of signing the Union of Utrecht in 1579.
- Note 2: A seignory comes closest to the concept of a *heerlijkheid*; there is no equivalent in English for the Dutch-language term. In its earliest history, Mechelen was a *heerlijkheid* of the Bishopric (later Prince-Bishopric) of Liège that exercised its rights through the Chapter of Saint Rumbold though at the same time the Lords of Berthout and later the Dukes of Brabant also exercised or claimed separate feudal rights.

French_Revolution

The French Revolution	
 The storming of the Bastille, 14 July 1789	
Participants	French society
Location	France
Date	1789–1799
Result	• A cycle of royal power being limited by uneasy constitutional monarchy—then abolition and replacement of the French king, aristocracy and church with a radical, secular, democratic republic—in turn becoming more authoritarian, militaristic and property-based. • Radical social change to forms based on Enlightenment principles of citizenship and inalienable rights, as well as nationalism and democracy. • Rise of Napoleon Bonaparte • Armed conflicts with other European countries

The **French Revolution** (French: *Révolution française*; 1789–1799), was a period of radical social and political upheaval in France that had a major impact on France and indeed all of Europe. The absolute monarchy that had ruled France for centuries collapsed in three years. French society underwent an epic transformation as feudal, aristocratic and religious privileges evaporated under a sustained assault from radical left-wing political groups, masses on the streets, and peasants in the countryside. Old ideas about tradition and hierarchy – of monarchy, aristocracy and religious authority – were abruptly overthrown by new Enlightenment principles of equality, citizenship and inalienable rights.

The French Revolution began in 1789 with the convocation of the Estates-General in May. The first year of the Revolution saw members of the Third Estate proclaiming the Tennis Court Oath in June, the assault on the Bastille in July, the passage of the Declaration of the Rights of Man and of the Citizen in August, and an epic march on Versailles that forced the royal court back to Paris in October. The next few years were dominated by tensions between various liberal assemblies and a right-wing monarchy intent on thwarting major reforms.

A republic was proclaimed in September 1792 and King Louis XVI was executed the next year. External threats also played a dominant role in the development of the Revolution. The French Revolutionary Wars started in 1792 and ultimately featured spectacular French victories that facilitated the conquest of the Italian Peninsula, the Low Countries and most territories west of the Rhine – achievements that had defied previous French governments for centuries.

Internally, popular sentiments radicalized the Revolution significantly, culminating in the rise of Maximilien Robespierre and the Jacobins and virtual dictatorship by the Committee of Public Safety during the Reign of Terror from 1793 until 1794 during which between 16,000 and 40,000 people were killed.[1] After the fall of the Jacobins and the execution of Robespierre, the Directory assumed control of the French state in 1795 and held power until 1799, when it was replaced by the Consulate under Napoleon Bonaparte.

After the Napoleonic Wars and ensuing rise and fall of Napoleon's First French Empire, a restoration of absolutist monarchy was followed by two further successful smaller revolutions (1830 and 1848). This meant the 19th century and process of modern France taking shape saw France again successively governed by a similar cycle of constitutional monarchy (1830–48), fragile republic (Second Republic) (1848–1852), and empire (Second Empire) (1852–1870). The modern era has unfolded in the shadow of the French Revolution. The growth of republics and liberal democracies, the spread of secularism, the development of modern ideologies and the invention of total war[2] all mark their birth during the Revolution.

Causes

The government of King Louis XVI of France faced a fiscal crisis in the 1780s.

Adherents of most historical models identify many of the same features of the *Ancien Régime* as being among the causes of the Revolution. Economic factors included hunger and malnutrition in the most destitute segments of the population, due to rising bread prices (from a normal 8 sous for a four-pound loaf to 12 sous by the end of 1789),[3] after several years of poor grain harvests. Bad harvests (caused in part by extreme weather from El Niño along with volcanic activity at Laki and Grímsvötn in 1783–1784), rising food prices, and an inadequate transportation system that hindered the shipment of bulk foods from rural areas to large population centers contributed greatly to the destabilization of French society in the years leading up to the Revolution.

Another cause was the state's effective bankruptcy due to the enormous cost of previous wars, particularly the financial strain caused by French participation in the American Revolutionary War. The national debt amounted to some 1,000–2,000 million livres. The social burdens caused by war included the huge war debt, made worse by the loss of France's colonial possessions in North America and the growing commercial dominance of Great Britain. France's inefficient and antiquated financial system was unable to manage the national debt, something which was both partially caused and exacerbated by the burden of an inadequate system of taxation. To obtain new money to head off default on the government's loans, the king called an Assembly of Notables in 1787.

Meanwhile, the royal court at Versailles was seen as being isolated from, and indifferent to, the hardships of the lower classes. While in theory King Louis XVI was an absolute monarch, in practice he was often indecisive and known to back down when faced with strong opposition. While he did reduce government expenditures, opponents in the parlements successfully thwarted his attempts at enacting much needed reforms. Those who were opposed to Louis' policies further undermined royal authority by distributing pamphlets (often reporting false or exaggerated information) that criticized the government and its officials, stirring up public opinion against the monarchy.[4]

Many other factors involved resentments and aspirations given focus by the rise of Enlightenment ideals. These included resentment of royal absolutism; resentment by peasants, laborers and the bourgeoisie toward the traditional seigneurial privileges possessed by the nobility; resentment of the Church's influence over public policy and institutions; aspirations for freedom of religion; resentment of aristocratic bishops by the poorer rural clergy; aspirations for social, political and economic equality, and (especially as the Revolution progressed) republicanism;

hatred of Queen Marie-Antoinette, who was falsely accused of being a spendthrift and an Austrian spy; and anger toward the King for firing finance minister Jacques Necker, among others, who were popularly seen as representatives of the people.[5]

Pre-revolution

Financial crisis

Caricature of the Third Estate carrying the First Estate (clergy) and the Second Estate (nobility) on its back.

Louis XVI ascended to the throne amidst a financial crisis; the state was nearing bankruptcy and outlays outpaced income.[6] This was because of France's financial obligations stemming from involvement in the Seven Years War and its participation in the American Revolutionary War.[7] In May 1776, finance minister Turgot was dismissed, after he failed to enact reforms. The next year, Jacques Necker, a foreigner, was appointed Comptroller-General of Finance. He could not be made an official minister because he was a Protestant.[8]

Necker realized that the country's extremely regressive tax system subjected the lower classes to a heavy burden,[8] while numerous exemptions existed for the nobility and clergy.[9] He argued that the country could not be taxed higher; that tax exemptions for the nobility and clergy must be reduced; and proposed that borrowing more money would solve the country's fiscal shortages. Necker published a report to support this claim that underestimated the deficit by roughly 36 million livres, and proposed restricting the power of the *parlements*.[8]

This was not received well by the King's ministers and Necker, hoping to bolster his position, argued to be made a minister. The King refused, Necker was fired, and Charles Alexandre de Calonne was appointed to the Comptrollership.[8] Calonne initially spent liberally, but he quickly realized the critical financial situation and proposed a new tax code.[10]

The proposal included a consistent land tax, which would include taxation of the nobility and clergy. Faced with opposition from the parlements, Calonne organised the summoning of the Assembly of Notables. But the Assembly failed to endorse Calonne's proposals and instead weakened his position through its criticism. In response, the King announced the calling of the Estates-General for May 1789, the first time the body had been summoned since 1614. This was a signal that the Bourbon monarchy was in a weakened state and subject to the demands of its people.[11]

Estates-General of 1789

The Estates-General was organized into three estates: the clergy, the nobility, and the rest of France.[12] On the last occasion that the Estates-General had met, in 1614, each estate held one vote, and any two could override the third. The *Parlement* of Paris feared the government would attempt to gerrymander an assembly to rig the results. Thus, they required that the Estates be arranged as in 1614.[13] The 1614 rules differed from practices of local assemblies, where each member had one vote and third estate membership was doubled. For example, in the Dauphiné the provincial assembly agreed to double the number of members of the third estate, hold membership elections, and allow one vote per member, rather than one vote per estate.[14]

The "Committee of Thirty," a body of liberal Parisians, began to agitate against voting by estate. This group, largely composed of the wealthy, argued for the Estates-General to assume the voting mechanisms of Dauphiné. They argued that ancient precedent was not sufficient, because "the people were sovereign."[15] Necker convened a Second

Assembly of Notables, which rejected the notion of double representation by a vote of 111 to 333.[15] The King, however, agreed to the proposition on 27 December; but he left discussion of the weight of each vote to the Estates-General itself.[16]

Elections were held in the spring of 1789; suffrage requirements for the Third Estate were for French-born or naturalised males only, at least 25 years of age, who resided where the vote was to take place and who paid taxes.

> *Pour être électeur du tiers état, il faut avoir 25 ans, être français ou naturalisé, être domicilié au lieu de vote et compris au rôle des impositions.*[17]

Strong turnout produced 1,201 delegates, including: "291 nobles, 300 clergy, and 610 members of the Third Estate."[16] To lead delegates, "Books of grievances" (*cahiers de doléances*) were compiled to list problems.[12] The books articulated ideas which would have seemed radical only months before; however, most supported the monarchical system in general. Many assumed the Estates-General would approve future taxes, and Enlightenment ideals were relatively rare.[13] [18]

Pamphlets by liberal nobles and clergy became widespread after the lifting of press censorship.[15] The Abbé Sieyès, a theorist and Catholic clergyman, argued the paramount importance of the Third Estate in the pamphlet *Qu'est-ce que le tiers état?* ("What is the Third Estate?"), published in January 1789. He asserted: "What is the Third Estate? Everything. What has it been until now in the political order? Nothing. What does it want to be? Something."[13] [19]

The meeting of the Estates General on 5 May 1789 in Versailles.

The Estates-General convened in the Grands Salles des Menus-Plaisirs in Versailles on 5 May 1789 and opened with a three-hour speech by Necker. The Third Estate demanded that the verification of deputies' credentials should be undertaken in common by all deputies, rather than each estate verifying the credentials of its own members internally; negotiations with the other estates failed to achieve this.[18] The commoners appealed to the clergy who replied they required more time. Necker asserted that each estate verify credentials and "the king was to act as arbitrator."[20] Negotiations with the other two estates to achieve this, however, were unsuccessful.[21]

National Assembly (1789)

The National Assembly taking the Tennis Court Oath (sketch by Jacques-Louis David).

On 10 June 1789, Abbé Sieyès moved that the Third Estate, now meeting as the *Communes* (English: "Commons"), proceed with verification of its own powers and invite the other two estates to take part, but not to wait for them. They proceeded to do so two days later, completing the process on 17 June.[22] Then they voted a measure far more radical, declaring themselves the National Assembly, an assembly not of the Estates but of "the People." They invited the other orders to join them, but made it clear they intended to conduct the nation's affairs with or without them.[23]

In an attempt to keep control of the process and prevent the Assembly from convening, Louis XVI ordered the closure of the Salle des États where the Assembly met, making an excuse that the carpenters needed to prepare the hall for a royal speech in two days. Weather did not allow an outdoor meeting, so the Assembly moved their deliberations to a nearby indoor real tennis court, where they proceeded to swear the Tennis Court Oath (20 June 1789), under which they agreed not to separate until they had given France a constitution.[24]

A majority of the representatives of the clergy soon joined them, as did 47 members of the nobility. By 27 June, the royal party had overtly given in, although the military began to arrive in large numbers around Paris and Versailles. Messages of support for the Assembly poured in from Paris and other French cities.[24]

National Constituent Assembly (1789–1791)

Storming of the Bastille

By this time, Necker had earned the enmity of many members of the French court for his overt manipulation of public opinion. Marie Antoinette, the King's younger brother the Comte d'Artois, and other conservative members of the King's privy council urged him to dismiss Necker as financial advisor. On 11 July 1789, after Necker published an inaccurate account of the government's debts and made it available to the public, the King fired him, and completely restructured the finance ministry at the same time.[25]

Many Parisians presumed Louis's actions to be aimed against the Assembly and began open rebellion when they heard the news the next day. They were also afraid that arriving soldiers – mostly foreign mercenaries – had been summoned to shut down the National Constituent Assembly. The Assembly, meeting at Versailles, went into nonstop session to prevent another eviction from their meeting place. Paris was soon consumed by riots, chaos, and widespread looting. The mobs soon had the support of some of the French Guard, who were armed and trained soldiers.[26]

On 14 July, the insurgents set their eyes on the large weapons and ammunition cache inside the Bastille fortress, which was also perceived to be a symbol of royal power. After several hours of combat, the prison fell that afternoon. Despite ordering a cease fire, which prevented a mutual massacre, Governor Marquis Bernard de Launay was beaten, stabbed and decapitated; his head was placed on a pike and paraded about the city. Although the fortress had held only seven prisoners (four forgers, two noblemen kept for immoral behavior, and a murder suspect), the Bastille served as a potent symbol of everything hated under the *Ancien Régime*. Returning to the Hôtel de Ville (city hall), the mob accused the *prévôt des marchands* (roughly, mayor) Jacques de Flesselles of treachery and butchered him.[27]

The Declaration of the Rights of Man and of the Citizen of 26 August 1789

The King, alarmed by the violence, backed down, at least for the time being. The Marquis de la Fayette took up command of the National Guard at Paris. Jean-Sylvain Bailly, president of the Assembly at the time of the Tennis Court Oath, became the city's mayor under a new governmental structure known as the *commune*. The King visited Paris, where, on 17 July he accepted a tricolore cockade, to cries of *Vive la Nation* ("Long live the Nation") and *Vive le Roi* ("Long live the King").[28]

Necker was recalled to power, but his triumph was short-lived. An astute financier but a less astute politician, Necker overplayed his hand by demanding and obtaining a general amnesty, losing much of the people's favour.

As civil authority rapidly deteriorated, with random acts of violence and theft breaking out across the country, members of the nobility, fearing for their safety, fled to neighboring countries; many of these *émigrés*, as they were called, funded counter-revolutionary causes within France and urged foreign monarchs to offer military support to a counter-revolution.[29]

By late July, the spirit of popular sovereignty had spread throughout France. In rural areas, many commoners began to form militias and arm themselves against a foreign invasion: some attacked the châteaux of the nobility as part of a general agrarian insurrection known as *"la Grande Peur"* ("the Great Fear"). In addition, wild rumours and paranoia caused widespread unrest and civil disturbances that contributed to the collapse of law and order.[30]

Working toward a constitution

On 4 August 1789, the National Constituent Assembly abolished feudalism (although at that point there had been sufficient peasant revolts to almost end feudalism already), in what is known as the August Decrees, sweeping away both the seigneurial rights of the Second Estate and the tithes gathered by the First Estate. In the course of a few hours, nobles, clergy, towns, provinces, companies and cities lost their special privileges.

On 26 August 1789, the Assembly published the Declaration of the Rights of Man and of the Citizen, which comprised a statement of principles rather than a constitution with legal effect. The National Constituent Assembly functioned not only as a legislature, but also as a body to draft a new constitution.

Necker, Mounier, Lally-Tollendal and others argued unsuccessfully for a senate, with members appointed by the crown on the nomination of the people. The bulk of the nobles argued for an aristocratic upper house elected by the nobles. The popular party carried the day: France would have a single, unicameral assembly. The King retained only a "suspensive veto"; he could delay the implementation of a law, but not block it absolutely. The Assembly eventually replaced the historic provinces with 83 *départements,* uniformly administered and roughly equal in area and population.

Amid the Assembly's preoccupation with constitutional affairs, the financial crisis had continued largely unaddressed, and the deficit had only increased. Honoré Mirabeau now led the move to address this matter, and the Assembly gave Necker complete financial dictatorship.

Women's March on Versailles

Fueled by rumors of a reception for the King's bodyguards on 1 October 1789 at which the national cockade had been trampled upon, on 5 October 1789 crowds of women began to assemble at Parisian markets. The women first marched to the Hôtel de Ville, demanding that city officials address their concerns.[31] The women were responding to the harsh economic situations they faced, especially bread shortages. They also demanded an end to royal efforts to block the National Assembly, and for the King and his administration to move to Paris as a sign of good faith in addressing the widespread poverty.

Engraving of the Women's March on Versailles, 5 October 1789

Getting unsatisfactory responses from city officials, as many as 7,000 women joined the march to Versailles, bringing with them cannons and a variety of smaller weapons. Twenty thousand National Guardsmen under the command of La Fayette responded to keep order, and members of the mob stormed the palace, killing several guards.[32] La Fayette ultimately persuaded the king to accede to the demand of the crowd that the monarchy relocate to Paris.

On 6 October 1789, the King and the royal family moved from Versailles to Paris under the "protection" of the National Guards, thus legitimizing the National Assembly.

Revolution and the Church

In this caricature, monks and nuns enjoy their new freedom after the decree of 16 February 1790

The Revolution caused a massive shift of power from the Roman Catholic Church to the state. Under the *Ancien Régime*, the Church had been the largest single landowner in the country, owning about 10% of the land in the kingdom.[33] The Church was exempt from paying taxes to the government, while it levied a tithe—a 10% tax on income, often collected in the form of crops—on the general population, which it then redistributed to the poor.[33] The power and wealth of the Church was highly resented by some groups. A small minority of Protestants living in France, such as the Huguenots, wanted an anti-Catholic regime and revenge against the clergy who discriminated against them. Enlightenment thinkers such as Voltaire helped fuel this resentment by denigrating the Catholic Church and destabilizing the French monarchy.[34] As historian John McManners argues, "In eighteenth-century France throne and altar were commonly spoken of as in close alliance; their simultaneous collapse ... would one day provide the final proof of their interdependence."[35]

This resentment toward the Church weakened its power during the opening of the Estates General in May 1789. The Church composed the First Estate with 130,000 members of the clergy. When the National Assembly was later created in June 1789 by the Third Estate, the clergy voted to join them, which perpetuated the destruction of the Estates General as a governing body.[36] The National Assembly began to enact social and economic reform. Legislation sanctioned on 4 August 1789 abolished the Church's authority to impose the tithe. In an attempt to address the financial crisis, the Assembly declared, on 2 November 1789, that the property of the Church was "at the disposal of the nation."[37] They used this property to back a new currency, the assignats. Thus, the nation had now also taken on the responsibility of the Church, which included paying the clergy, caring for the poor, the sick and the orphaned.[38] In December, the Assembly began to sell the lands to the highest bidder to raise revenue, effectively decreasing the value of the assignats by 25% in two years.[39] In autumn 1789, legislation abolished monastic vows and on 13 February 1790 all religious orders were dissolved.[40] Monks and nuns were encouraged to return to private life and a small percentage did eventually marry.[41]

The Civil Constitution of the Clergy, passed on 12 July 1790, turned the remaining clergy into employees of the state. This established an election system for parish priests and bishops and set a pay rate for the clergy. Many Catholics objected to the election system because it effectively denied the authority of the Pope in Rome over the French Church. Eventually, in November 1790, the National Assembly began to require an oath of loyalty to the Civil Constitution from all the members of the clergy.[41] This led to a schism between those clergy who swore the required oath and accepted the new arrangement and those who remained loyal to the Pope. Overall, 24% of the clergy nationwide took the oath.[42] Widespread refusal led to legislation against the clergy, "forcing them into exile, deporting them forcibly, or executing them as traitors."[39] Pope Pius VI never accepted the Civil Constitution of the Clergy, further isolating the Church in France. During the Reign of Terror, extreme efforts of de-Christianization ensued, including the imprisonment and massacre of priests and destruction of churches and religious images throughout France. An effort was made to replace the Catholic Church altogether, with civic festivals replacing religious ones. The establishment of the Cult of Reason was the final step of radical de-Christianization. These events led to a widespread disillusionment with the Revolution and to counter-rebellions across France. Locals often resisted de-Christianization by attacking revolutionary agents and hiding members of the clergy who were being hunted. Eventually, Robespierre and the Committee of Public Safety were forced to denounce the campaign,[43] replacing the Cult of Reason with the deist but still non-Christian Cult of the Supreme Being. The Concordat of 1801 between Napoleon and the Church ended the de-Christianization period and established the rules for a relationship between the Catholic Church and the French State that lasted until it was abrogated by the Third Republic via the separation of church and state on 11 December 1905. The persecution of the Church led to a counter-revolution known as the Revolt in the Vendée, whose suppression is considered by some to be the first modern genocide.

Intrigues and radicalism

Factions within the Assembly began to clarify. The aristocrat Jacques Antoine Marie de Cazalès and the abbé Jean-Sifrein Maury led what would become known as the right wing, the opposition to revolution (this party sat on the right-hand side of the Assembly). The "Royalist democrats" or *monarchiens*, allied with Necker, inclined toward organising France along lines similar to the British constitutional model; they included Jean Joseph Mounier, the Comte de Lally-Tollendal, the comte de Clermont-Tonnerre, and Pierre Victor Malouet, comte de Virieu.

The "National Party", representing the centre or centre-left of the assembly, included Honoré Mirabeau, La Fayette, and Bailly; while Adrien Duport, Barnave and Alexandre Lameth represented somewhat more extreme views. Almost alone in his radicalism on the left was the Arras lawyer Maximilien Robespierre. Abbé Sieyès led in proposing legislation in this period and successfully forged consensus for some time between the political centre and the left. In Paris, various committees, the mayor, the assembly of representatives, and the individual districts each claimed authority independent of the others. The increasingly middle-class National Guard under La Fayette also slowly emerged as a power in its own right, as did other self-generated assemblies.

The *Fête de la Fédération* on 14 July 1790 celebrated the establishment of the constitutional monarchy

The Assembly abolished the symbolic paraphernalia of the *Ancien Régime*— armorial bearings, liveries, etc. – which further alienated the more conservative nobles, and added to the ranks of the *émigrés*. On 14 July 1790, and for several days following, crowds in the Champ de Mars celebrated the anniversary of the fall of the Bastille with the *Fête de la Fédération*; Talleyrand performed a mass; participants swore an oath of "fidelity to the nation, the law, and the king"; the King and the royal family actively participated.[44]

The electors had originally chosen the members of the Estates-General to serve for a single year. However, by the terms of the Tennis Court Oath, the *communes* had bound themselves to meet continuously until France had a constitution. Right-wing elements now argued for a new election, but Mirabeau prevailed, asserting that the status of the assembly had fundamentally changed, and that no new election should take place before completing the constitution.

In late 1790, the French army was in considerable disarray. The military officer corps was largely composed of noblemen, who found it increasingly difficult to maintain order within the ranks. In some cases, soldiers (drawn from the lower classes) had turned against their aristocratic commanders and attacked them. At Nancy, General Bouillé successfully put down one such rebellion, only to be accused of being anti-revolutionary for doing so. This and other such incidents spurred a mass desertion as more and more officers defected to other countries, leaving a dearth of experienced leadership within the army.[45]

This period also saw the rise of the political "clubs" in French politics. Foremost among these was the Jacobin Club; 152 members had affiliated with the Jacobins by 10 August 1790. The Jacobin Society began as a broad, general organization for political debate, but as it grew in members, various factions developed with widely differing views. Several of these fractions broke off to form their own clubs, such as the Club of '89.[46]

Meanwhile, the Assembly continued to work on developing a constitution. A new judicial organisation made all magistracies temporary and independent of the throne. The legislators abolished hereditary offices, except for the monarchy itself. Jury trials started for criminal cases. The King would have the unique power to propose war, with the legislature then deciding whether to declare war. The Assembly abolished all internal trade barriers and suppressed guilds, masterships, and workers' organisations: any individual gained the right to practice a trade through the purchase of a license; strikes became illegal.[47]

In the winter of 1791, the Assembly considered, for the first time, legislation against the *émigrés*. The debate pitted the safety of the Revolution against the liberty of individuals to leave. Mirabeau prevailed against the measure, which he referred to as "worthy of being placed in the code of Draco".[45] But Mirabeau died on 2 April 1791 and, before the end of the year, the new Legislative Assembly adopted this draconian measure.[48]

Royal flight to Varennes

Louis XVI, egged on by Marie Antoinette and other members of his family, opposed the course of the Revolution, but rejected the potentially treacherous aid of the other monarchs of Europe. He cast his lot with General Bouillé, who condemned both the emigration and the Assembly, and promised him refuge and support in his camp at Montmédy. On the night of 20 June 1791, the royal family fled the Tuileries Palace dressed as servants, while their servants dressed as nobles.

The return of the royal family to Paris on 25 June 1791, after their failed flight to Varennes

However, late the next day, the King was recognised and arrested at Varennes (in the Meuse *département*). He and his family were brought back to Paris under guard, still dressed as servants. Pétion, Latour-Maubourg, and Antoine Pierre Joseph Marie Barnave, representing the Assembly, met the royal family at Épernay and returned with them. From this time, Barnave became a counselor and supporter of the royal family. When they returned to Paris, the crowd greeted them in silence. The Assembly provisionally suspended the King. He and Queen Marie Antoinette remained held under guard.[49] [50] [51] [52] [53]

Completing the constitution

As most of the Assembly still favoured a constitutional monarchy rather than a republic, the various groups reached a compromise which left Louis XVI as little more than a figurehead: he was forced to swear an oath to the constitution, and a decree declared that retracting the oath, heading an army for the purpose of making war upon the nation, or permitting anyone to do so in his name would amount to abdication.[54]

However, Jacques Pierre Brissot drafted a petition, insisting that in the eyes of the nation Louis XVI was deposed since his flight. An immense crowd gathered in the Champ de Mars to sign the petition. Georges Danton and Camille Desmoulins gave fiery speeches. The Assembly called for the municipal authorities to "preserve public order". The National Guard under La Fayette's command confronted the crowd. The soldiers responded to a barrage of stones by firing into the crowd, killing between 13 and 50 people.[55]

In the wake of this massacre the authorities closed many of the patriotic clubs, as well as radical newspapers such as Jean-Paul Marat's *L'Ami du Peuple*. Danton fled to England; Desmoulins and Marat went into hiding.

Meanwhile, a new threat arose from abroad: Holy Roman Emperor Leopold II, Frederick William II of Prussia, and the King's brother Charles-Philippe, comte d'Artois, issued the Declaration of Pillnitz, which considered the cause of Louis XVI as their own, demanded his absolute liberty and implied an invasion of France on his behalf if the revolutionary authorities refused its conditions.[56] The French people expressed no respect for the dictates of foreign monarchs, and the threat of force merely hastened their militarisation.[57]

Even before the "Flight to Varennes", the Assembly members had determined to debar themselves from the legislature that would succeed them, the Legislative Assembly. They now gathered the various constitutional laws they had passed into a single constitution, showed remarkable strength in choosing not to use this as an occasion for major revisions, and submitted it to the recently restored Louis XVI, who accepted it, writing "I engage to maintain it at home, to defend it from all attacks from abroad, and to cause its execution by all the means it places at my disposal". The King addressed the Assembly and received enthusiastic applause from members and spectators. With this capstone, the National Constituent Assembly adjourned in a final session on 30 September 1791.[58]

Mignet argued that the "constitution of 1791... was the work of the middle class, then the strongest; for, as is well known, the predominant force ever takes possession of institutions... In this constitution the people was the source of all powers, but it exercised none."[54]

Legislative Assembly (1791–1792)

Failure of the constitutional monarchy

Under the Constitution of 1791, France would function as a constitutional monarchy. The King had to share power with the elected Legislative Assembly, but he still retained his royal veto and the ability to select ministers. The Legislative Assembly first met on 1 October 1791, and degenerated into chaos less than a year later. In the words of the 1911 Encyclopædia Britannica: "In the attempt to govern, the Assembly failed altogether. It left behind an empty treasury, an undisciplined army and navy, and a people debauched by safe and successful riot."[59] The Legislative Assembly consisted of about 165 Feuillants (constitutional monarchists) on the right, about 330 Girondists (liberal republicans) and Jacobins (radical revolutionaries) on the left, and about 250 deputies unaffiliated with either faction. Early on, the King vetoed legislation that threatened the *émigrés* with death and that decreed that every non-juring clergyman must take within eight days the civic oath mandated by the Civil Constitution of the Clergy. Over the course of a year, such disagreements would lead to a constitutional crisis.

Constitutional crisis

On the night of 10 August 1792, insurgents and popular militias, supported by the revolutionary Paris Commune, assailed the Tuileries Palace and massacred the Swiss Guards who were assigned for the protection of the king. The royal family ended up prisoners and a rump session of the Legislative Assembly suspended the monarchy; little more than a third of the deputies were present, almost all of them Jacobins.[60]

On 10 August 1792 the Paris Commune stormed the Tuileries Palace and massacred the Swiss Guards

What remained of a national government depended on the support of the insurrectionary Commune. The Commune sent gangs into the prisons to try arbitrarily and butcher 1400 victims, and addressed a circular letter to the other cities of France inviting them to follow this example. The Assembly could offer only feeble resistance. This situation persisted until the Convention, elected by universal male suffrage and charged with writing a new constitution, met on 20 September 1792 and became the new *de facto* government of France. The next day it abolished the monarchy and declared a republic. This date was later retroactively adopted as the beginning of Year One of the French Republican Calendar.

War and Counter-Revolution (1792–1797)

The politics of the period inevitably drove France towards war with Austria and its allies. The King, many of the Feuillants and the Girondins specifically wanted to wage war. The King (and many Feuillants with him) expected war would increase his personal popularity; he also foresaw an opportunity to exploit any defeat: either result would make him stronger. The Girondins wanted to export the Revolution throughout Europe and, by extension, to defend the Revolution within France. The forces opposing war were much weaker. Barnave and his supporters among the Feuillants feared a war they thought France had little chance to win and which they feared might lead to greater radicalization of the revolution. On the other end of the political spectrum Robespierre opposed a war on two grounds, fearing that it would strengthen the monarchy and military at the expense of the revolution, and that it would incur the anger of ordinary people in Austria and elsewhere. The Austrian emperor Leopold II, brother of Marie Antoinette, may have wished to avoid war, but he died on 1 March 1792.[61] France preemptively declared war on Austria (20 April 1792) and Prussia joined on the Austrian side a few weeks later. The invading Prussian army faced little resistance until checked at the Battle of Valmy (20 September 1792), and forced to withdraw.

The new-born Republic followed up on this success with a series of victories in Belgium and the Rhineland in the fall of 1792. The French armies defeated the Austrians at the Battle of Jemappes on 6 November, and had soon taken over most of the Austrian Netherlands. This brought them into conflict with Britain and the Dutch Republic, which wished to preserve the independence of the southern Netherlands from France. After the king's execution in January 1793, these powers, along with Spain and most other European states, joined the war against France. Almost immediately, French forces faced defeat on many fronts, and were driven out of their newly conquered territories in the spring of 1793. At the same time, the republican regime was forced to deal with rebellions against its authority in much of western and southern France. But the allies failed to take advantage of French disunity, and by the autumn of 1793 the republican regime had defeated most of the internal rebellions and halted the allied advance into France itself.

The stalemate was broken in the summer of 1794 with dramatic French victories. They defeated the allied army at the Battle of Fleurus, leading to a full Allied withdrawal from the Austrian Netherlands. They followed up by a campaign which swept the allies to the east bank of the Rhine and left the French, by the beginning of 1795, conquering Holland itself. The House of Orange was expelled and replaced by the Batavian Republic, a French satellite state. These victories led to the collapse of the coalition against France. Prussia, having effectively abandoned the coalition in the fall of 1794, made peace with revolutionary France at Basel in April 1795, and soon thereafter Spain, too, made peace with France. Of the major powers, only Britain and Austria remained at war with France.

It was during this time that *La Marseillaise* was first sung. Originally titled *Chant de guerre pour l'Armée du Rhin* ("War Song for the Army of the Rhine"), the song was written and composed by Claude Joseph Rouget de Lisle in 1792. It was adopted in 1795 as the nation's first anthem.

National Convention (1792–1795)

Execution of Louis XVI

In the Brunswick Manifesto, the Imperial and Prussian armies threatened retaliation on the French population if it were to resist their advance or the reinstatement of the monarchy. This among other things made Louis appear to be conspiring with the enemies of France. On 17 January 1793 Louis was condemned to death for "conspiracy against the public liberty and the general safety" by a close majority in Convention: 361 voted to execute the king, 288 voted against, and another 72 voted to execute him subject to a variety of delaying conditions. The former Louis XVI, now simply named *Citoyen Louis Capet* (Citizen Louis Capet), was executed by guillotine on 21 January 1793 on the *Place de la Révolution*, former *Place Louis XV*, now called the Place de la Concorde.[62] Royalty across Europe was horrified and many heretofore neutral countries soon joined the war against revolutionary France.

Execution of Louis XVI in what is now the Place de la Concorde, facing the empty pedestal where the statue of his grandfather, Louis XV, had stood.

Economy

When war went badly, prices rose and the *sans-culottes* — poor labourers and radical Jacobins – rioted; counter-revolutionary activities began in some regions. This encouraged the Jacobins to seize power through a parliamentary *coup*, backed up by force effected by mobilising public support against the Girondist faction, and by utilising the mob power of the Parisian *sans-culottes*. An alliance of Jacobin and *sans-culottes* elements thus became the effective centre of the new government. Policy became considerably more radical, as "The Law of the Maximum" set food prices and led to executions of offenders.[63] This policy of price control was coeval with the Committee of Public Safety's rise to power and the Reign of Terror. The Committee first attempted to set the price for only a limited number of grain products but, by September 1793, it expanded the "maximum" to cover all foodstuffs and a long list of other goods.[64] Widespread shortages and famine ensued. The Committee reacted by sending dragoons into the countryside to arrest farmers and seize crops. This temporarily solved the problem in Paris, but the rest of the country suffered. By the spring of 1794, forced collection of food was not sufficient to feed even Paris and the days of the Committee were numbered. When Robespierre went to the guillotine in July of that year the crowd jeered, "There goes the dirty maximum!"[65]

Reign of Terror

The Committee of Public Safety came under the control of Maximilien Robespierre, a lawyer, and the Jacobins unleashed the Reign of Terror (1793–1794). According to archival records, at least 16,594 people died under the guillotine or otherwise after accusations of counter-revolutionary activities.[66] A number of historians note that as many as 40,000 accused prisoners may have been summarily executed without trial or died awaiting trial.[66] [67]

Satirical cartoon from England lampooning the excesses of the Revolution as symbolized through the guillotine: between 18,000 and 40,000 people were executed during the Reign of Terror

On 2 June 1793, Paris sections — encouraged by the *enragés* ("enraged ones") Jacques Roux and Jacques Hébert – took over the Convention, calling for administrative and political purges, a low fixed price for bread, and a limitation of the electoral franchise to *sans-culottes* alone.[68] With the backing of the National Guard, they managed to persuade the Convention to arrest 31 Girondin leaders, including Jacques Pierre Brissot. Following these arrests, the Jacobins gained control of the Committee of Public Safety on 10 June, installing the *revolutionary dictatorship*. On 13 July, the assassination of Jean-Paul Marat — a Jacobin leader and journalist known for his bloodthirsty rhetoric — by Charlotte Corday, a Girondin, resulted in further increase of Jacobin political influence. Georges Danton, the leader of the August 1792 uprising against the King, undermined by several political reversals, was removed from the Committee and Robespierre, "the Incorruptible", became its most influential member as it moved to take radical measures against the Revolution's domestic and foreign enemies.[69]

Meanwhile, on 24 June, the Convention adopted the first republican constitution of France, variously referred to as the French Constitution of 1793 or Constitution of the Year I. It was progressive and radical in several respects, in particular by establishing universal male suffrage. It was ratified by public referendum, but normal legal processes were suspended before it could take effect.[70]

Queen Marie Antoinette on the way to the guillotine on 16 October 1793 (drawing by Jacques-Louis David)

War in the Vendée

The War in the Vendée was a royalist uprising that was suppressed by the republican forces in 1796

In Vendée, peasants revolted against the French Revolutionary government in 1793. They resented the changes imposed on the Roman Catholic Church by the Civil Constitution of the Clergy (1790) and broke into open revolt in defiance of the Revolutionary government's military conscription.[71] This became a guerrilla war, known as the War in the Vendée.[72] North of the Loire, similar revolts were started by the so-called Chouans (royalist rebels).[73]

After the defeat at Savenay, when regular warfare in the Vendée was at an end, the French general Francois Joseph Westermann is argued by some historians to have penned a letter (its veracity is disputed)[74] [75] to the Committee of Public Safety, stating:

> "There is no more Vendée. It died with its wives and its children by our free sabres. I have just buried it in the woods and the swamps of Savenay. According to the orders that you gave me, I crushed the children under the feet of the horses, massacred the women who, at least for these, will not give birth to any more brigands. I do not have a prisoner to reproach me. I have exterminated all. The roads are sown with corpses. At Savenay, brigands are arriving all the time claiming to surrender, and we are shooting them non-stop... Mercy is not a revolutionary sentiment."[76] [77]

Other historians doubt the authenticity of this document and point out that the claims in it were patently false — there were in fact thousands of living Vendean prisoners, the revolt had been far from crushed, and the Convention had explicitly decreed that women, children and unarmed men were to be treated humanely.[78] It has been hypothesized that if the letter is authentic, Westermann may have been attempting to exaggerate the intensity of his actions and his success, because he was eager to avoid being purged for his opposition to *sans-culotte* generals (he was later guillotined together with Danton's group).[79]

The revolt and its suppression, including both combat casualties and massacres and executions on both sides, are thought to have taken between 117,000 and 250,000 lives (170,000 according to the latest estimates).[80] Because of the extremely brutal forms that the Republican repression took in many places, certain historians such as Reynald Secher have called the event a "genocide". This description has become popular in the mass media,[81] but has largely been rejected by mainstream scholars.[82]

Facing local revolts and foreign invasions in both the East and West of the country, the most urgent government business was the war. On 17 August, the Convention voted for general conscription, the *levée en masse*, which mobilized all citizens to serve as soldiers or suppliers in the war effort.

The National Convention subsequently enacted more legislation, voting on 9 September to establish *sans-culottes* paramilitary forces, *revolutionary armies*, and to force farmers to surrender grain demanded by the government. On 17 September, the *Law of Suspects* was passed, which authorized the charging of counter-revolutionaries with "crimes against liberty." On 29 September, the Convention extended price limits from grain and bread to other household goods and established the Law of the Maximum, intended to prevent price gouging and supply food to the cities.[83]

The guillotine as a symbol

The guillotine became the symbol of a string of executions. Louis XVI had already been guillotined before the start of the terror; Queen Marie Antoinette, Barnave, Bailly, Brissot and other leading Girondins, Philippe Égalité (despite his vote for the death of the King), Madame Roland and many others were executed by guillotine. The Revolutionary Tribunal summarily condemned thousands of people to death by the guillotine, while mobs beat other victims to death.

The Festival of the Supreme Being on 8 June 1794

At the peak of the terror, the slightest hint of counter-revolutionary thoughts or activities (or, as in the case of Jacques Hébert, revolutionary zeal exceeding that of those in power) could place one under suspicion, and trials did not always proceed according to contemporary standards of due process. Sometimes people died for their political opinions or actions, but many for little reason beyond mere suspicion, or because some others had a stake in getting rid of them. Most of the victims received an unceremonious trip to the guillotine in an open wooden cart (the tumbrel). In the rebellious provinces, the government representatives had unlimited authority and some engaged in extreme repressions and abuses. For example, Jean-Baptiste Carrier became notorious for the *Noyades* ("drownings") he organized in Nantes;[84] his conduct was judged unacceptable even by the Jacobin government and he was recalled.[85]

Another anti-clerical uprising was made possible by the installment of the Republican Calendar on 24 October 1793. Against Robespierre's concepts of Deism and Virtue, Hébert's (and Chaumette's) atheist movement initiated a religious campaign to dechristianize society. The climax was reached with the celebration of the flame of Reason in Notre Dame Cathedral on 10 November.[86]

The Reign of Terror ultimately weakened the revolutionary government, while temporarily ending internal opposition. The Jacobins expanded the size of the army, and Carnot replaced many aristocratic officers with soldiers who had demonstrated their patriotism, if not their ability. The Republican army was able to throw back the Austrians, Prussians, British, and Spanish. At the end of 1793, the army began to prevail and revolts were defeated with ease. The Ventôse Decrees (February–March 1794) proposed the confiscation of the goods of exiles and opponents of the Revolution, and their redistribution to the needy. However this policy was never fully implemented.[87]

In the spring of 1794, both extremist *enragés* such as Hébert and moderate Montagnard *indulgents* such as Danton were charged with counter-revolutionary activities, tried and guillotined. On 7 June Robespierre, who had previously condemned the *Cult of Reason*, advocated a new state religion and recommended the Convention acknowledge the existence of the "Supreme Being".[88]

Thermidorian Reaction

On 27 July 1794, the Thermidorian Reaction led to the arrest and execution of Robespierre, Louis de Saint-Just, and other leading Jacobins. The new government was predominantly made up of Girondists who had survived the Terror, and after taking power, they took revenge as well by persecuting even those Jacobins who had helped to overthrow Robespierre, banning the Jacobin Club, and executing many of its former members in what was known as the White Terror.[89] [90]

The execution of Robespierre on 28 July 1794 marked the end of the Reign of Terror

In the wake of excesses of the Terror, the Convention approved the new "Constitution of the Year III" on 22 August 1795. A French plebiscite ratified the document, with about 1,057,000 votes for the constitution and 49,000 against.[91] The results of the voting were announced on 23 September 1795, and the new constitution took effect on 27 September 1795.[91]

The Constitutional Republic: The Directory (1795–1799)

The new constitution created the *Directoire* (English: Directory) and the first bicameral legislature in French history.[92] The parliament consisted of two houses: the *Conseil des Cinq-Cents* (Council of the Five Hundred), with 500 representatives, and the *Conseil des Anciens* (Council of Elders), with 250 senators. Executive power went to five "directors," named annually by the *Conseil des Anciens* from a list submitted by the *Conseil des Cinq-Cents*. Furthermore, the universal male suffrage of 1793 was replaced by limited suffrage based on property.[93]

With the establishment of the Directory, contemporary observers might have assumed that the Revolution was finished. Citizens of the war-weary nation wanted stability, peace, and an end to conditions that at times bordered on chaos. Those who wished to restore the monarchy and the *Ancien Régime* by putting Louis XVIII on the throne, and those who would have renewed the Reign of Terror were insignificant in number. The possibility of foreign interference had vanished with the failure of the First Coalition. The earlier atrocities had made confidence or goodwill between parties impossible. The same instinct of self-preservation which had led the members of the Convention to claim so large a part in the new legislature and the whole of the Directory impelled them to keep their predominance. However, many French citizens distrusted the Directory,[94] and the directors could achieve their purposes only by extraordinary means. They habitually disregarded the terms of the constitution, and, even when the elections that they rigged went against them, the directors routinely used draconian police measures to quell dissent. Moreover, to prolong their power the directors were driven to rely on the military, which desired war and grew less and less civic-minded.[95]

Other reasons influenced them in the direction of war. State finances during the earlier phases of the Revolution had been so thoroughly ruined that the government could not have met its expenses without the plunder and the tribute of foreign countries. If peace were made, the armies would return home and the directors would have to face the exasperation of the rank-and-file who had lost their livelihood, as well as the ambition of generals who could, in a moment, brush them aside. Barras and Rewbell were notoriously corrupt themselves and screened corruption in others. The patronage of the directors was ill-bestowed, and the general maladministration heightened their unpopularity.[96]

Napoléon Bonaparte in the *coup d'état* of 18 Brumaire, which marked the end of the revolution

The constitutional party in the legislature desired toleration of the nonjuring clergy, the repeal of the laws against the relatives of the émigrés, and some merciful discrimination toward the émigrés themselves. The directors baffled all such endeavours. On the other hand, the socialist conspiracy of Babeuf was easily quelled. Little was done to improve the finances, and the assignats continued to fall in value.

The new régime met opposition from remaining Jacobins and the royalists. The army suppressed riots and counter-revolutionary activities. In this way the army and its successful general, Napoleon Bonaparte eventually gained total power.

On 9 November 1799 (18 Brumaire of the Year VIII) Napoleon Bonaparte staged the *coup of 18 Brumaire* which installed the Consulate. This effectively led to Bonaparte's dictatorship and eventually (in 1804) to his proclamation as *Empereur* (emperor), which brought to a close the specifically republican phase of the French Revolution.[97]

Symbolism in the French Revolution

Early depiction of the tricolour in the hands of a *sans-culotte* during the French Revolution

The French Revolution was a time of upheaval, especially towards traditional ideology, in almost every sense: the current monarch, King Louis XVI, was executed; the Catholic Church was all but abolished; a new calendar was created; and a new Republican government was established. In order to effectively illustrate the differences between the new Republic and the old regime, the leaders needed to implement a new set of symbols to be celebrated instead of the old religious and monarchical symbolism. To this end, symbols were borrowed from historic cultures and redefined, while those of the old regime were either destroyed or reattributed acceptable characteristics. These revised symbols were used to instill in the public a new sense of tradition and reverence for the Enlightenment and the Republic.[98]

Fasces

Fasces, likes many other symbols of the French Revolution, are Roman in origin. Fasces are a bundle of birch rods containing an axe. In Roman times, the fasces symbolized the power of magistrates who could order the beating of a criminal, representing union and accord with the Roman Republic.[98] The French Republic continued this Roman symbol to represent state power, justice, and unity. During the French Revolution the fasces image is seen in conjunction with many other symbols. This is seen with many emblems of the French Revolution. Though seen throughout the French Revolution, perhaps the most well known French reincarnation of the fasces is the Fasces surmounted by a Phrygian cap. This image has no display of an axe or a strong central state; rather, it symbolizes the power of the liberated people by placing the Liberty Cap on top of the classical symbol of power.[98]

Liberty cap

The Liberty cap, also known as the Phrygian cap, or pileus, is a brimless, felt cap that is conical in shape with the tip pulled forward. The cap was originally worn by ancient Romans and Greeks.[99] The cap implies ennobling effects, as seen in its association with Homer's Ulysses and the mythical twins, Castor and Pollux. The emblem's popularity during the French Revolution is due in part to its importance in ancient Rome: its use alludes to the Roman ritual of manumission of slaves, in which a freed slave receives the bonnet as a symbol of his newfound liberty. The Roman tribune Lucius Appuleius Saturninus incited the slaves to insurrection by displaying a pileus as if it were a standard.[100] The pileus cap is often red in color. This type of cap was worn by revolutionaries at the fall of the Bastille. According to the Revolutions de Paris, it became "the symbol of the liberation from all servitudes, the sign for unification of all the enemies of despotism."[98] The pileus competed with the Phrygian cap, a similar cap that covered the ears and the nape of the neck, for popularity. The Phrygian cap eventually supplanted the pileus and usurped its symbolism, becoming synonymous with republican liberty.[98]

Liberty Tree

The Liberty Tree, officially adopted in 1792, is a symbol of the everlasting Republic, national freedom, and political revolution.[98] It has historic roots in revolutionary France as well as America, as a symbol that was shared by the two nascent republics.[101] The tree was chosen as a symbol of the French Revolution because it is a symbol of fertility in French folklore,[102] which provided a simple transition from revering it for one reason to another. The American colonies also used the idea of a Liberty Tree to celebrate their own acts of insurrection against the British, starting with the Stamp Act riot in 1765.[103] The riot culminated in the hanging in effigy of two Stamp Act politicians on a large elm tree. The elm tree began to be celebrated as a symbol of Liberty in the American colonies.[103] It was adopted as a symbol that needed to be living and growing, along with the Republic. To that end, the tree is portrayed as a sapling, usually of an oak tree in French interpretation.[104] The Liberty Tree serves as a constant celebration of the spirit of political freedom.

Hercules

The symbol of Hercules was first adopted by the Old Regime to represent the monarchy.[105] Hercules was an ancient Greek hero who symbolized strength and power. The symbol was used to represent the sovereign authority of the King over France during the reign of the Bourbon monarchs.[106] However, the monarchy was not the only ruling power in French history to use the symbol of Hercules to declare its power.

During the Revolution, the symbol of Hercules was revived to represent nascent revolutionary ideals. The first use of Hercules as a revolutionary symbol was during a festival celebrating the National Assembly's victory over federalism on 10 August 1793.[107] This Festival of Unity consisted of four stations around Paris which featured symbols representing major events of the Revolution which embodied revolutionary ideals of liberty, unity, and power.[108] The statue of Hercules, placed at the station commemorating the fall of Louis XVI, symbolized the power of the French people over their former oppressors. The statue's foot was placed on the throat of the Hydra, which represented the tyranny of federalism which the new Republic had vanquished.[107] In one hand, the statue grasped a club, a symbol of power, while in the other grasping the fasces which symbolized the unity of the French people.[109] The image of Hercules assisted the new Republic in establishing its new Republican moral system.[108] Hercules thus evolved from a symbol of the sovereignty of the monarch into a symbol of the new sovereign authority in France: the French people.[110] This transition was made easily for two reasons. First, because Hercules was a famous mythological figure, and had previously been used by the monarchy, he was easily recognized by educated French observers.[106] It was not necessary for the revolutionary government to educate the French people on the background of the symbol. Additionally, Hercules recalled the classical age of the Greeks and the Romans, a period which the revolutionaries identified with republican and democratic ideals. These connotations made Hercules an easy choice to represent the powerful new sovereign people of France.

During the more radical phase of the Revolution from 1793 to 1794, the usage and depiction of Hercules changed. These changes to the symbol were due to revolutionary leaders believing the symbol was inciting violence among the common citizens.[111] The triumphant battles of Hercules and the overcoming of enemies of the Republic became less prominent. In discussions over what symbol to use for the Seal of the Republic, the image of Hercules was considered but eventually ruled out in favor of Marianne.[111] Hercules was on the coin of the Republic.[111] However, this Hercules was not the same image as that of the pre-Terror phases of the Revolution. The new image of Hercules was more domesticated. He appeared more paternal, older, and wiser, rather than the warrior-like images in the early stages of the French Revolution.[111] Unlike his 24 foot statue in the Festival of the Supreme Being, he was now the same size as Liberty and Equality.[111] Also the language on the coin with Hercules was far different than the rhetoric of pre-revolutionary depictions. On the coins the words, "uniting Liberty and Equality" were used.[111] This is opposed to the forceful language of early Revolutionary rhetoric and rhetoric of the Bourbon monarchy. By 1798, the Council of Ancients had discussed the "inevitable" change from the problematic image of Hercules, and Hercules was eventually phased out in favor of an even more docile image.[111]

Role of women

Women had no political rights in pre-Revolutionary France; they could not vote or hold any political office. They were considered "passive" citizens; forced to rely on men to determine what was best for them in the government. It was the men who defined these categories, and women were forced to accept male domination in the political sphere.[112] The *Encyclopédie*, published by a group of philosophers over the years 1751–1777, summarized French male beliefs of women. A woman was a "failed man," the fetus not fully developed in the womb. "Women's testimony is in general light and subject to variation; this is why it is taken more seriously than that of men" as opposed to men, upon whom "Nature seems to have conferred... the right to govern." In general, "men are more capable than women of ably governing particular matters".[113] Instead, women were taught to be committed to their husbands and "all his interests... [to show] attention and care... [and] sincere and discreet zeal for his salvation." A woman's education often consisted of learning to be a good wife and mother; as a result women were not supposed to be involved in the political sphere, as the limit of their influence was the raising of future citizens.[114]

When the Revolution opened, some women struck forcefully, using the volatile political climate to assert their active natures. In the time of the Revolution, women could not be kept out of the political sphere; they swore oaths of loyalty, "solemn declarations of patriotic allegiance, [and] affirmations of the political responsibilities of citizenship." Throughout the Revolution, women such as Pauline Léon and her Society of Revolutionary Republican Women fought for the right to bear arms, used armed force and rioted.[115]

Even before Léon, some liberals had advocated equal rights for women including women's suffrage. Nicolas de Condorcet was especially noted for his advocacy, in his articles published in the *Journal de la Société de 1789*, and by publishing *De l'admission des femmes au droit de cité* ("For the Admission to the Rights of Citizenship For Women") [116] in 1790.

Feminist agitation

The March to Versailles is but one example of feminist militant activism during the French Revolution. While largely left out of the thrust for increasing rights of citizens, as the question was left indeterminate in the Declaration of the Rights of Man,[117] activists such as Pauline Léon and Théroigne de Méricourt agitated for full citizenship for women.[118] Women were, nonetheless, "denied political rights of 'active citizenship' (1791) and democratic citizenship (1793)."[117]

Pauline Léon, on 6 March 1792, submitted a petition signed by 319 women to the National Assembly requesting permission to form a garde national in order to defend Paris in case of military invasion.[118] Léon requested permission be granted to women to arm themselves with pikes, pistols, sabers and rifles, as well as the privilege of drilling under the French Guards. Her request was denied.[119] Later in 1792, Théroigne de Méricourt made a call for the creation of "legions of amazons" in order to protect the revolution. As part of her call, she claimed that the right to bear arm would transform women into citizens.[120]

On 20 June 1792 a number of armed women took part in a procession that "passed through the halls of the Legislative Assembly, into the Tuilleries Gardens, and then through the King's residence."[121] Militant women also assumed a special role in the funeral of Marat, following his murder on 13 July 1793. As part of the funeral procession, they carried the bathtub in which Marat had been murdered as well as a shirt stained with Marat's blood.[122]

The most radical militant feminist activism was practiced by the Society of Revolutionary Republican Women, which was founded by Léon and her colleague, Claire Lacombe on 10 May 1793.[123] The goal of the club was "to deliberate on the means of frustrating the projects of the enemies of the Republic." Up to 180 women attended the meetings of the Society.[124] Of special interest to the Society was "combating hoarding [of grain and other staples] and inflation."[125]

Later, on 20 May 1793, women were at the fore of a crowd that demanded "bread and the Constitution of 1793."[126] When their cries went unnoticed, the women went on a rampage, "sacking shops, seizing grain and kidnapping officials."[127]

Most of these outwardly activist women were punished for their actions. The kind of punishment received during the Revolution included public denouncement, arrest, execution, or exile. Théroigne de Méricourt was arrested, publicly flogged and then spent the rest of her life sentenced to an insane asylum. Pauline Léon and Claire Lacombe were arrested, later released, and continued to receive ridicule and abuse for their activism. Many of the women of the Revolution were even publicly executed for "conspiring against the unity and the indivisibility of the Republic".[128]

These are but a few examples of the militant feminism that was prevalent during the French Revolution. While little progress was made toward gender equality during the Revolution, the activism of French feminists was bold and particularly significant in Paris.

Women writers

Olympe de Gouges was the author of the *Declaration of the Rights of Woman and the Female Citizen* in 1791

While some women chose a militant, and often violent, path, others chose to influence events through writing, publications, and meetings. Olympe de Gouges wrote a number of plays, short stories, and novels. Her publications emphasized that women and men are different, but this shouldn't stop them from equality under the law. In her "Declaration on the Rights of Woman" she insisted that women deserved rights, especially in areas concerning them directly, such as divorce and recognition of illegitimate children. De Gouges also expressed non-gender political views; even before the start of the terror, Olympe de Gouges addressed Robespierre using the pseudonym "Polyme" calling him the Revolution's "infamy and shame." She warned of the Revolution's building extremism saying that leaders were "preparing new shackles if [the French people's liberty were to] waver." Stating that she was willing to sacrifice herself by jumping into the Seine if Robespierre were to join her, de Gouges desperately attempted to grab the attention of the French citizenry and alert them to the dangers that Robespierre embodied.[129] In addition to these bold writings, her defense of the king was one of the factors leading to her execution. An influential figure, one of her suggestions early in the Revolution, to have a voluntary, patriotic tax, was adopted by the National Convention in 1789.[130]

Madame Roland (aka Manon or Marie Roland) was another important female activist. Her political focus was not specifically on women or their liberation. She focused on other aspects of the government, but was a feminist by virtue of the fact that she was a woman working to influence the world. Her personal letters to leaders of the Revolution influenced policy; in addition, she often hosted political gatherings of the Brissotins, a political group which allowed women to join. While limited by her gender, Madame Roland took it upon herself to spread Revolutionary ideology and spread word of events, as well as to assist in formulating the policies of her political allies. Though unable to directly write policies or carry them through to the government, Roland was able to influence her political allies and thus promote her political agenda. Roland attributed women's lack of education to the public view that women were too weak or vain to be involved in the serious business of politics. She believed that it was this inferior education that turned them into foolish people, but women "could easily be concentrated and solidified upon objects of great significance" if given the chance.[131] As she was led to the scaffold, Madame Roland shouted "O liberty! What crimes are committed in thy name!" Witnesses of her life and death, editors, and readers helped to finish her writings and several editions were published posthumously. While she did not focus on gender politics in her writings, by taking an active role in the tumultuous time of the Revolution, Roland took a stand for women of the time and proved they could take an intelligent active role in politics.[132]

Though women did not gain the right to vote as a result of the Revolution, they still greatly expanded their political participation and involvement in governing. They set precedents for generations of feminists to come.

Counter-revolutionary women

A major aspect of the French Revolution was the dechristianisation movement, a movement that many common people did not agree with. Especially for women living in rural areas of France, the demise of the Catholic Church meant a loss of normalcy. For instance, the ringing of Church bells resonating through the town called people to confession and was a symbol of unity for the community.[133] With the onset of the dechristianisation campaign the Republic silenced these bells and sought simultaneously to silence the religious fervor of the majority Catholic population.[133] When these revolutionary changes to the Church were implemented, it spawned a counter-revolutionary movement, particularly amongst women. Although some of these women embraced the political and social amendments of the Revolution, they opposed the dissolution of the Catholic Church and the formation of revolutionary cults like the Cult of the Supreme Being advocated by Robespierre.[134] As Olwen Hufton argues, these women began to see themselves as the "defenders of faith".[135] They took it upon themselves to protect the Church from what they saw as a heretical change to their faith, enforced by revolutionaries.

Counter-revolutionary women resisted what they saw as the intrusion of the state into their lives.[136] Economically, many peasant women refused to sell their goods for assignats because this form of currency was unstable and was backed by the sale of confiscated Church property.[135] By far the most important issue to counter-revolutionary women was the passage and the enforcement of the Civil Constitution of the Clergy in 1790. In response to this measure, women in many areas began circulating anti-oath pamphlets and refused to attend masses held by priests who had sworn oaths of loyalty to the Republic.[136] This diminished the social and political influence of the juring priests because they presided over smaller congregations and counter-revolutionary women did not seek them for baptisms, marriages or confession.[137] Instead, they secretly hid nonjuring priests and attended clandestine traditional masses.[138] These women continued to adhere to traditional practices such as Christian burials and naming their children after saints in spite of revolutionary decrees to the contrary.[139]

It was this determined resistance to the Civil Constitution of the Clergy and the dechristianisation campaigns that played a major role in the re-emergence of the Catholic Church as a prominent social institution. In fact, Olwen Hufton notes about the Counter-Revolutionary women: "for it is her commitment to her religion which determines in the post-Thermidorean period the re-emergence of the Catholic Church...".[140] Although they struggled, these women were eventually vindicated in their bid to reestablish the Church and thereby also to reestablish traditional family life and social stability.[141] This was seen in the Concordat of 1801, which formally reinstated the Catholic

Church in France.[142] This act came after years of failed attempts at dechristianisation or state-controlled religion, which were thwarted in part due to the resistance of devout counter-revolutionary women. After the upheaval of the revolutionary period, the reestablishment of the Church was seen by many people as a welcome return to normalcy.

Legacy

The French Revolution has received enormous amounts of historical attention, both from the general public and from scholars and academics. The views of historians, in particular, have been characterized as falling along ideological lines, with disagreement over the significance and the major developments of the Revolution.[143] Alexis de Tocqueville argued that the Revolution was a manifestation of a more prosperous middle class becoming conscious of its social importance.[144] Other thinkers, like the conservative Edmund Burke, maintained that the Revolution was the product of a few conspiratorial individuals who brainwashed the masses into subverting the old order—a claim rooted in the belief that the revolutionaries had no legitimate complaints.[145] Other historians, influenced by Marxist thinking, have emphasized the importance of the peasants and the urban workers in presenting the Revolution as a gigantic class struggle.[146] In general, scholarship on the French Revolution initially studied the political ideas and developments of the era, but it has gradually shifted towards social history that analyzes the impact of the Revolution on individual lives.[147]

Historians widely regard the Revolution as one of the most important events in human history, and the end of the early modern period, which started around 1500, is traditionally attributed to the onset of the French Revolution in 1789.[148] The Revolution is, in fact, often seen as marking the "dawn of the modern era".[149] Within France itself, the Revolution permanently crippled the power of the aristocracy and drained the wealth of the Church, although the two institutions survived despite the damage they sustained. After the collapse of the First Empire in 1815, the French public lost the rights and privileges earned since the Revolution, but they remembered the participatory politics that characterized the period, with one historian commenting: "Thousands of men and even many women gained firsthand experience in the political arena: they talked, read, and listened in new ways; they voted; they joined new organizations; and they marched for their political goals. Revolution became a tradition, and republicanism an enduring option."[150] Some historians argue that the French people underwent a fundamental transformation in self-identity, evidenced by the elimination of privileges and their replacement by rights as well as the growing decline in social deference that highlighted the principle of equality throughout the Revolution.[151] The Revolution represented the most significant and dramatic challenge to political absolutism up to that point in history and, despite its failures, spread democratic ideals throughout Europe and ultimately the world.[152] It had a profound impact on the Russian Revolution and its ideas inspired Mao Zedong in his efforts at constructing a communist state in China.[153]

See also

- Biens nationaux
- Dual revolution
- History of democracy
- La Révolution française (film)
- List of people granted honorary French citizenship during the French Revolution
- List of revolutions and rebellions
- Military career of Napoleon Bonaparte
- Napoleonic code
- Jean-Nicolas Pache
- Dechristianisation of France during the French Revolution
- Rise of nationalism in Europe

Other revolutions or rebellions in French history

- Camisard Rebellion, French Huguenots (1710–1715)
- Haitian Revolution, Haiti colony (1791–1804)
- July Revolution (1830)
- Canut revolts of the July Monarchy
- French Revolution of 1848
- Resistance to the coup of 1851
- Paris Commune of 1871
- French Army Mutinies (1917)
- French Resistance during World War II
- May 1968 in France, a noteworthy rebellion, though not quite a revolution

Notes

[1] Donald Greer, *The Incidence of the Terror during the French Revolution: A Statistical Interpretation* (1935).

[2] Bell, David Avrom (2007). *The First Total War: Napoleon's Europe and the birth of warfare as we know it.* New York: Houghton Mifflin Harcourt. p. 51. ISBN 0618349650. "The French Revolution, which began in 1789 and led to the total war of 1792–1815...."

[3] Hibbert. Pg 96.

[4] "Encyclopædia Britannica — Traite" (http://www.britannica.com/EBchecked/topic/602094/traite). . Retrieved 16 October 2008.

[5] William Doyle, *The Oxford History of the French Revolution* (2nd ed. 2003), pp.73–74

[6] Frey, p. 3

[7] "France's Financial Crisis: 1783–1788" (http://www.sparknotes.com/history/european/frenchrev/section1.html). . Retrieved 26 October 2008.

[8] Hibbert, p. 35, 36

[9] Frey, p. 2

[10] Doyle, *The French Revolution: A very short introduction*, p. 34

[11] Doyle 2003, p. 93

[12] Frey, pp. 4, 5

[13] Doyle 2001, p. 38

[14] Doyle 1989, p.89

[15] Neely, p. 56

[16] Hibbert, pp.42–45

[17] Assemblée Nationale (French) (http://www.assemblee-nationale.fr/histoire/suffrage_universel/suffrage-1789.asp)

[18] Neely, pp. 63, 65

[19] Furet, p. 45

[20] Hibbert, p. 54

[21] Schama 2004, p.300–301

[22] John Hall Stewart. *A Documentary Survey of the French Revolution*. New York: Macmillan, 1951, p. 86.

[23] Schama 2004, p.303

[24] Schama 2004, p.312

[25] Schama 2004, p.317

[26] Schama 2004, p.331

[27] Schama 2004, p.344

[28] Schama 2004, p.357

[29] Lefebvre, pp.187–188.

[30] Hibbert, 93

[31] Doyle 1989, p.121

[32] Doyle 1989, p.122

[33] Censer and Hunt, Liberty, Equality, Fraternity: Exploring the French Revolution, 4.

[34] Censer and Hunt, Liberty, Equality, Fraternity: Exploring the French Revolution, 16.

[35] John McManners, The French Revolution and the Church, 5.

[36] John McManners, The French Revolution and the Church, 50, 4.

[37] National Assembly legislation cited in John McManners, The French Revolution and the Church, 27.

[38] John McManners, The French Revolution and the Church, 27.

[39] Censer and Hunt, Liberty, Equality, Fraternity: Exploring the French Revolution, 61.

[40] Emmet Kennedy, A Cultural History of the French Revolution, 148.

[41] Censer and Hunt, Liberty, Equality, Fraternity: Exploring the French Revolution, 92.
[42] Emmet Kennedy, A Cultural History of the French Revolution, 151.
[43] Censer and Hunt, Liberty, Equality, Fraternity: Exploring the French Revolution, 92–94.
[44] Schama 2004, p.433–434
[45] Mignet, François (1824). *Histoire de la Révolution française*. Chapter III.
[46] Schama 2004, p.449
[47] Schama 2004, p.442
[48] Schama 2004, p.496
[49] Encyclopædia Britannica Eleventh Edition
[50] Lindqvist, Herman (1991). Axel von Fersen. Stockholm: Fischer & Co
[51] Loomis, Stanley (1972). The Fatal Friendship. Avon Books — ISBN 0-931933-33-1
[52] Timothy Tackett, When the King Took Flight (Cambridge: Harvard University Press, 2003)
[53] The Flight to Varennes • Memoir by the Duchesse d'Angoulême (http://penelope.uchicago.edu/Thayer/E/Gazetteer/Places/Europe/France/_Texts/CROROY/Fuite_de_Varennes*.html)
[54] Mignet, François (1824). *Histoire de la Révolution française*. Chapter IV.
[55] Schama 2004, p.481
[56] Schama 2004, p.500
[57] Soboul (1975), pp. 226–227.
[58] Lefebvre, p. 212.
[59] "French Revolution" (http://www.1911encyclopedia.org/French_Revolution). About LoveToKnow 1911. . Retrieved 10 April 2009.
[60] Pfeiffer, L. B., "The Uprising of June 20, 1792," p.221. New Era Printing Company, Lincoln: 1913."
[61] Schama 2004, p.505
[62] Doyle 2002, p. 196
[63] White, E. "The French Revolution and the Politics of Government Finance, 1770–1815." *The Journal of Economic History* 1995, p 244
[64] Schuettinger, Robert. "Forty Centuries of Wage and Price Controls." Heritage Foundation, 2009. p. 45
[65] Bourne, Henry. "Maximum Prices in France." American Historical Review, October 1917, p. 112.
[66] Gough, Hugh (1998). *The Terror in the French Revolution*. p. 77.
[67] Doyle 1989, p. 258
[68] Schama 2004, p.616
[69] Schama 2004, p.641
[70] Schama 2004, p.637
[71] In a Corner of France, Long Live the Old Regime (http://query.nytimes.com/gst/fullpage.html?res=950DE2DB1231F934A25755C0A96F948260&sec=&spon=&pagewanted=all), New York Times
[72] McPhee, Peter Review of Reynald Secher, A French Genocide: The Vendée (http://www.h-france.net/vol4reviews/mcphee3.html) H-France Review Vol. 4 (March 2004), No. 26
[73] Hibbert, p. 321.
[74] Frédéric Augris, Henri Forestier, général à 18 ans, Éditions du Choletais, 1996
[75] Jean-Clément Martin, Contre-Révolution, Révolution et Nation en France, 1789–1799, éditions du Seuil, collection Points, 1998, p. 219
[76] Davies, Norman. *Europe: A history* Pimlico, (1997). p. 705
[77] Schama 2004, p.666
[78] Jean-Clément Martin, Guerre de Vendée, dans l'Encyclopédie Bordas, Histoire de la France et des Français, Paris, Éditions Bordas, 1999, p 2084, et Contre-Révolution, Révolution et Nation en France, 1789–1799, p.218.
[79] Jean-Clément Martin, *Violence et Révolution. Essai sur la naissance d'un mythe national*, éditions du Seuil, 2006, p. 181
[80] 117,000 according to Reynald Secher, *La Vendée-Vengé, le Génocide franco-français* (1986); 200,000–250,000 according to Jean-Clément Martin, *La Vendée et la France*, Éditions du Seuil, collection Points, 1987; 200,000 according to Louis-Marie Clénet, *La Contre-révolution*, Paris, PUF, collection Que sais-je?, 1992; 170,000 according to Jacques Hussenet (dir.), *« Détruisez la Vendée ! » Regards croisés sur les victimes et destructions de la guerre de Vendée*, La Roche-sur-Yon, Centre vendéen de recherches historiques, 2007, p.148.
[81] In a Corner of France, Long Live the Old Regime (http://query.nytimes.com/gst/fullpage.html?res=950DE2DB1231F934A25755C0A96F948260&sec=&spon=&pagewanted=all). *The New York Times.* 17 June 1989
[82] Michel Vovelle, « L'historiographie de la Révolution Française à la veille du bicentenaire », *Estudos avançados*, octobre-décembre 1987, volume 1, n° 1, p. 61–72. (http://www.scielo.br/scielo.php?pid=S0103-40141987000100006&script=sci_arttext) ou (http://www.scielo.br/pdf/ea/v1n1/v1n1a06.pdf)
[83] Schama 2004, p.646
[84] Jean-Baptiste Carrier (http://www.britannica.com/EBchecked/topic/97009/Jean-Baptiste-Carrier), *Encyclopædia Britannica*
[85] Soboul (1975), p. 384.
[86] Schama 2004, p.658
[87] Schama 2004, p.689
[88] Schama 2004, p.706
[89] Soboul (1975), pp. 425–428.

[90] Schama, p.852.
[91] Doyle 1989, p.320
[92] Cole et al 1989, p.39
[93] Doyle, *Oxford History* (2003) pp 318–40
[94] Doyle, *Oxford History* p.331
[95] Doyle, *Oxford History*, p.332
[96] Doyle, *Oxford History*, pp. 322–23
[97] David Nicholls, *Napoleon: a biographical companion* (1999) p.
[98] Censer and Hunt, "How to Read Images" LEF CD-ROM
[99] Encyclopædia Britannica entry (http://www.britannica.com/EBchecked/topic/460407/pileus)
[100] Harden, "Liberty Caps and Liberty Trees" 90
[101] Harden, "Liberty Bells and Liberty Trees" 75–77
[102] Ozouf, "Festivals and the French Revolution" 123
[103] Harden, "Liberty Bells and Liberty Trees" 75
[104] Harden, "Liberty Bells and Liberty Trees" 88
[105] Hunt 1984, 89.
[106] Hunt 1984, 101–102
[107] Hunt 1984, 96
[108] Censer and Hunt 2001, 92
[109] Hunt 1984, 97
[110] Hunt 1984, 103
[111] Hunt 1984, 113
[112] Scott "Only Paradoxes to Offer" 34–35
[113] Encyclopedia "Woman"
[114] Marquise de Maintenon, "Writings" 321
[115] Dalton "Madame Roland" 262
[116] http://www.pinn.net/~sunshine/book-sum/condorcet4.html
[117] Rebel Daughters: Women and the French Revolution Edited by Sara E Melzer and Leslie W. Rabine pg. 79
[118] Women and the Limits of Citizenship in the French Revolution by Olwen W. Hufton pg. 23–24
[119] Rebel Daughters by Sara E Melzer and Leslie W. Rabine pg. 89
[120] Women and the Limits of Citizenship by Olwen W. Hufton pg. 23–24
[121] Rebel Daughters by Sara E Melzer and Leslie W. Rabine pg. 91
[122] Women and the Limits of Citizenship by Olwen W. Hufton pg. 31
[123] Rebel Daughters by Sara E Melzer and Leslie W. Rabine pg. 92
[124] Deviant Women of the French Revolution and the Rise of Feminism by Lisa Beckstrand pg. 17
[125] Women and the Limits of Citizenship by Olwen W. Hufton pg. 25
[126] Gender, Society and Politics: France and Women 1789–1914 by James H. McMillan pg. 24
[127] Gender, Society and Politics by McMillan pg. 24
[128] Deviant Women by Beckstrand pg. 20
[129] De Gouges "Writings" 564–568
[130] Mousset "Women's Rights" 49
[131] Dalton "Madame Roland" 262–267
[132] Walker "Virtue" 413–416
[133] Hufton, Olwen. Women and the Limits of Citizenship 1992 pg. 106–107
[134] Desan pg. 452
[135] Hufton, Olwen. "In Search of Counter-Revolutionary Women." 1998 pg. 303
[136] Hufton, Olwen. Women and the Limits of Citizenship 1992 pg. 104
[137] Hufton, Olwen. "In Search of Counter-Revolutionary Women." 1998 pg. 311
[138] Hufton, Olwen. Women and the Limits of Citizenship 1992 pg. 105
[139] Hufton, Olwen. "In Search of Counter-Revolutionary Women." 1998 pg. 304
[140] Hufton, Olwen. "In Search of Counter-Revolutionary Women." 1998 pg. 305
[141] Hufton, Olwen. Women and the Limits of Citizenship 1992 pg. 130
[142] Hufton, Olwen. "In Search of Counter-Revolutionary Women." 1998 pg. 326
[143] Rude p. 12-4
[144] Rude, p. 15
[145] Rude, p. 12
[146] Rude, p. 17
[147] Rude, p. 12-20
[148] Frey, Foreword

[149] Frey, Preface
[150] Hanson, p. 189
[151] Hanson, 191
[152] Riemer, Neal; Simon, Douglas (28 January 1997). *The New World of Politics: An Introduction to Political Science* (http://books.google.com/books?id=gKa3FTpH49oC&pg=PA106). Rowman & Littlefield. p. 106. ISBN 978-0-939693-41-2. . Retrieved 18 October 2011.
[153] Hanson, 193

References

- Carlyle, Thomas (2002) [1837]. *The French Revolution: A History* (http://books.google.com/?id=nmRdAAAAMAAJ&printsec=frontcover&dq=French+Revolution+carlyle). The Modern Library. ISBN 0375760229.
- Censer, Jack; Lynn Hunt (2001). *Liberty, Equality, Fraternity: Exploring the French Revolution*. Pennsylvania: Pennsylvania State University Press.
- Cole, Alistair; Peter Campbell (1989). *French electoral systems and elections since 1789* (http://books.google.com/?id=msOIAAAAMAAJ&q="first+bicameral+legislature"+directory+french&dq="first+bicameral+legislature"+directory+french). Gower.
- Dalton, Susan (2001). *"Gender and the Shifting Ground of Revolutionary Politics: The Case of Madame Roland" Canadian Journal of History*. ISSN 00084107.
- Desan, S. "The Role of Women in Religious Riots during the French Revolution." Eighteenth-Century Studies. Vol. 22, No. 3, Special Issue: The French Revolution in Culture (Spring, 1989), pp. 451–468.
- Doyle, William (1990). *The Oxford history of the French Revolution* (http://books.google.com/?id=N7chHQAACAAJ&dq=Oxford+history+of+the+French+Revolution) (3rd ed.). Oxford: Oxford University Press. ISBN 0192852213.
- Doyle, William (2001). *The French Revolution: A very short introduction* (http://books.google.com/?id=B-Zf4topzfQC&dq=0192853961). Oxford: Oxford University Press. ISBN 0192853961.
- Doyle, William (2002). *The Oxford history of the French Revolution* (http://books.google.com/?id=oMYWKlevAV4C) (2nd ed.). Oxford University Press. ISBN 019925298X.
- Frey, Linda; Marsha Frey (2004). *The French Revolution* (http://books.google.com/?id=RFgXl1EG5soC&dq=Short+History+of+the+French+Revolution+Doyle). Westport, Connecticut: Greenwood Press. ISBN 0313321930.
- Furet, Francois (1995). *Revolutionary France, 1770–1880* (http://books.google.com/?id=EPsEJIdFz3AC&dq=Revolutionary+France+Furet). Blackwell Publishing. ISBN 0631198083.
- Hanson, Paul (2009). *Contesting the French Revolution* (http://books.google.com/?id=dsCf8Q0xqO4C&pg=PA186&dq=legacy+french+revolution&cd=17#v=onepage&q=legacy french revolution). Blackwell Publishing. ISBN 9781405160834.
- Hibbert, Christopher (1980). *The Days of the French Revolution* (http://books.google.com/?id=yr1z38HCkdAC&dq=Hibbert+French+Revolution). New York: Quill, William Morrow. ISBN 0688037046.
- Hufton, Olwen (1992). *Women and the Limits of Citizenship in the French Revolution*. Toronto, Canada: University of Toronto Press.
- Hufton, Olwen. "In Search of Counter-Revolutionary Women." The French Revolution: Recent debates and New Controversies. Ed. Gary Kates. New York, NY. Routledge (1998).
- Hunt, Lynn (1984). *Politics, Culture, and Class in the French Revolution*. Berkley: University of California Press.
- Levy, Darline Gay and Harriet B. Applewhite. "Women and Militant Citizenship in Revolutionary Paris," in Rebel Daughters, ed. Sara e. Melzer and Leslie W. Rabine (New York: Oxford University Press 1992).
- Kennedy, Emmet (1989). *A Cultural History of the French Revolution*. New Haven: Yale University Press.
- Lefebvre, Georges (1971). *The French Revolution: From Its Origins to 1793* (http://books.google.com/?id=I2oA5MtB0ZkC&dq=Lefebvre+estates+general). Columbia University Press. ISBN 0231085982.

- Marquise de Maintenon "Instruction to the Nuns of St. Louis," in Writings by Pre-Revolutionary French Women. ed. Anne R. Larsen and Colette H Winn. (New York: Garland Publishing Inc., 2000), 321.
- McManners, John (1969). *The French Revolution and the Church.* New York: Harper and Row.
- Mousset, Sophie (2007). *Women's Rights and the French Revolution.* Joy Poirel. Transaction Publishers. ISBN 0765803453.
- Neely, Sylvia (2008). *A Concise History of the French Revolution* (http://books.google.com/?id=fccjTyOQiYwC&dq=The+Estates-General). Rowman & Littlefield. ISBN 0742534111.
- Rude, George (1991). *The French Revolution: Its Causes, Its History and Its Legacy After 200 Years* (http://books.google.com/?id=f1pMIbvzKckC&dq=causes+of+the+French+Revolution). Grove Press. ISBN 0802132723.
- Schama, Simon (2004) [1989]. *Citizens.* Penguin. ISBN 0141017279.
- Scott, Joan Wallach. "A Woman Who Has Only Paradoxes to Offer," in Rebel Daughters, ed. Sara e. Melzer and Leslie W. Rabine (New York: Oxford University Press 1992).
- Soboul, Albert (1975). *The French Revolution 1787–1799.* New York: Vintage. ISBN 039471220X.
- Soboul, Albert (1977). *A short history of the French Revolution: 1789–1799* (http://books.google.com/?id=GSdw2oC60cwC&dq=Soboul+French+Revolution). Geoffrey Symcox. University of California Press, Ltd. ISBN 0520034198.
- Walker, Leslie H. " Sweet and Consoling Virtue: The Memoirs of Madame Roland (http://0-www.jstor.org.bianca.penlib.du.edu/)" Eighteenth-Century Studies, French Revolutionary Culture (2001): 403–419.
- "Women." The Encyclopedia of Diderot and d'Alembert. University of Michigan Library, n.d. Web. 10/29/09. < http://quod.lib.umich.edu/d/did/>.
- This article incorporates text from a publication now in the public domain: Chisholm, Hugh, ed. (1911). *Encyclopædia Britannica* (11th ed.). Cambridge University Press.
- *This article incorporates text from the public domain* History of the French Revolution from 1789 to 1814 (http://www.gutenberg.org/dirs/etext06/7hfrr10.txt), *by François Mignet (1824), as made available by Project Gutenberg.*

Further reading

- Baker, Keith M. ed. *The French Revolution and the Creation of Modern Political Culture* (Oxford, 1987–94) *vol 1: The Political Culture of the Old Regime,* ed. K.M. Baker (1987); *vol. 2: The Political Culture of the French Revolution,* ed. C. Lucas (1988); *vol. 3: The Transformation of Political Culture, 1789–1848,* eds. F. Furet & M. Ozouf (1989); *vol. 4: The Terror,* ed. K.M. Baker (1994). excerpt and text search vol 4 (http://www.amazon.com/French-Revolution-Creation-Political-Culture/dp/0080413870/)
- Blanning, T.C.W. *The French Revolutionary Wars 1787–1802* (1996).
- Censer, Jack R. "Amalgamating the Social in the French Revolution." *Journal of Social History 2003* 37(1): 145–150. Issn: 0022-4529 Fulltext: in Project Muse and Ebsco
- Davies, Peter. *The French Revolution: A Beginner's Guide* (2009), 192pp
- Doyle, William. *The Oxford History of the French Revolution* (1989). online complete edition (http://www.questia.com/library/book/the-oxford-history-of-the-french-revolution-by-william-doyle.jsp); also excerpt and online search from Amazon.com (http://www.amazon.com/Oxford-History-French-Revolution/dp/019925298X/)
- Doyle, William. *The French Revolution: A Very Short Introduction.* (2001), 120pp; online edition (http://www.questia.com/library/book/the-french-revolution-a-very-short-introduction-by-william-doyle.jsp)
- Doyle, William. *Origins of the French Revolution* (3rd ed. 1999) online edition (http://www.questia.com/library/book/origins-of-the-french-revolution-by-william-doyle.jsp)
- Dunne, John. "Fifty Years of Rewriting the French Revolution: Signposts Main Landmarks and Current Directions in the Historiographical Debate," *History Review.* (1998) pp 8+ online edition (http://www.questia.

com/PM.qst?a=o&d=5001505046)

- Englund, Steven. *Napoleon: A Political Life.* (2004). 575 pages; the best political biography excerpt and text search (http://www.amazon.com/Napoleon-Political-Life-Steven-Englund/dp/0674018036/)
- Fremont-Barnes, Gregory. ed. *The Encyclopedia of the French Revolutionary and Napoleonic Wars: A Political, Social, and Military History* (ABC-CLIO: 3 vol 2006)
- Frey, Linda S. and Marsha L. Frey. *The French Revolution.* (2004) 190pp online edition (http://www.questia.com/library/book/the-french-revolution-by-linda-s-frey-marsha-l-frey.jsp)
- Furet, François. *The French Revolution, 1770–1814* (1996) excerpt and text search (http://www.amazon.com/French-Revolution-1770-1814-History-France/dp/0631202994/)
- Furet, François and Mona Ozouf, eds. *A Critical Dictionary of the French Revolution* (1989), 1120pp; long essays by scholars; conservative perspective; stress on history of ideas excerpt and online search from Amazon.com (http://www.amazon.com/Critical-Dictionary-French-Revolution/dp/0674177282/)
- Hufton, Olwen. *Women and the Limits of Citizenship in the French Revolution* University of Toronto Press (1992).
- Hunt, Lynn. *The Family Romance of the French Revolution.* (1992)
- Hunt, Lynn. "Politics, Culture, and Class in the French Revolution." (1984)
- Germani, Ian and Robin Swayles. *Symbols, myths and images of the French Revolution.* University of Regina Publications. 1998. ISBN 9780889771086
- Griffith, Paddy. *The Art of War of Revolutionary France 1789–1802,* (1998); 304 pp; excerpt and text search (http://www.amazon.com/Art-War-Revolutionary-France-1789-1802/dp/1853673358/)
- Jones, Colin. *The Longman Companion to the French Revolution* (1989)
- Jones, Colin. *The Great Nation: France from Louis XV to Napoleon* (2002) excerpt and text search (http://www.amazon.com/Great-Nation-Napoleon-Penguin-History/dp/0140130934/)
- Kaiser, Thomas, and Dale Van Kley. *From Deficit to Deluge: The Origins of the French Revolution* (2010)
- Kates, Gary. *The French Revolution* (2nd ed. 2005), 308pp; essays by scholars; excerpts and online search from Amazon.com (http://www.amazon.com/gp/reader/0415358329/)
- Lefebvre, Georges. *The French Revolution* (2 vol 1957) classic Marxist synthesis. complete online edition vol 1 (http://www.questia.com/library/book/the-french-revolution-vol-1-by-georges-lefebvre-elizabeth-moss-evanson.jsp); also excerpt and online search from Amazon.com (http://www.amazon.com/gp/reader/0231085982/)
- Neely, Sylvia. *A Concise History of the French Revolution* (2008)
- Palmer, Robert R. *The Age of the Democratic Revolution: A Political History of Europe and America, 1760–1800.* (2 vol 1959), highly influential comparative history; vol 1 online (http://www.questia.com/read/22790906)
- Paxton, John. *Companion to the French Revolution* (1987), hundreds of short entries.
- Rothenberg, Gunther E. "The Origins, Causes, and Extension of the Wars of the French Revolution and Napoleon," *Journal of Interdisciplinary History,* Vol. 18, No. 4, (Spring, 1988), pp. 771–793 in JSTOR (http://www.jstor.org/pss/204824)
- Rude, George F. and Harvey J. Kaye. *Revolutionary Europe, 1783–1815* (2000), scholarly survey excerpt and text search (http://www.amazon.com/Revolutionary-Europe-Blackwell-Classic-Histories/dp/0631221905/)
- Schroeder, Paul. *The Transformation of European Politics, 1763–1848.* 1996; Thorough coverage of diplomatic history; hostile to Napoleon; online edition (http://www.questia.com/library/book/the-transformation-of-european-politics-1763-1848-by-paul-w-schroeder.jsp)
- Schwab, Gail M., and John R. Jeanneney, eds. *The French Revolution of 1789 and Its Impact* (1995) online edition (http://www.questia.com/library/book/the-french-revolution-of-1789-and-its-impact-by-john-r-jeanneney-gail-m-schwab.jsp)

- Scott, Samuel F. and Barry Rothaus. *Historical Dictionary of the French Revolution, 1789–1799* (2 vol 1984), short essays by scholars
- Schama, Simon. *Citizens. A Chronicle of the French Revolution* (1989), highly readable narrative by scholar excerpt and text search (http://www.amazon.com/Citizens-Simon-Schama/dp/0141017279/)
- Sutherland, D.M.G. *France 1789–1815. Revolution and Counter-Revolution* (2nd ed. 2003, 430pp excerpts and online search from Amazon.com (http://www.amazon.com/French-Revolution-Empire-Quest-Civic/dp/0631233636/)
- Woshinsky, Barbara R. *Imaging Women's Conventual Spaces in France, 1600–1800: The Cloister Disclosed.* Burlington, Vermont. Ashgate (2010).

External links

- " French Revolution (http://www.newadvent.org/cathen/13009a.htm)" in the *Catholic Encyclopedia*
- Open University course (http://openlearn.open.ac.uk/course/view.php?id=1515)made)
- Entry on Encyclopedia.com (http://www.encyclopedia.com/doc/1E1-FrenchRe.html) from the Columbia Encyclopedia
- Primary source documents (http://www.fordham.edu/halsall/mod/modsbook13.html) from The Internet Modern History Sourcebook
- Liberty, Equality, Fraternity: Exploring the French Revolution (http://chnm.gmu.edu/revolution/), a collaborative site by the Center for History and New Media (George Mason University) and the American Social History Project (City University of New York)
- The Origins of the French Revolution (http://www.historyguide.org/intellect/lecture11a.html), The French Revolution: The Moderate Stage, 1789–1792 (http://www.historyguide.org/intellect/lecture12a.html), and The French Revolution: The Radical Stage, 1792–1794 (http://www.historyguide.org/intellect/lecture13a.html), three essays from The History Guide: Lectures on Modern European Intellectual History
- Vancea, S. The Cahiers de Doleances of 1789 (http://cliojournal.wikispaces.com/The+Cahiers+de+Doleances+of+1789), Clio History Journal, 2008.
- english.republique.de (http://english.republique.de/links.php) A large collection of links

Article Sources and Contributors

Colwich,_Staffordshire *Source*: http://en.wikipedia.org/w/index.php?title=Colwich%2C_Staffordshire *Contributors*: Cyan22, Diannaa, Handsaw, Lightmouse, Malcolma, Marek69, Peter morrell, Sjorford, Skinsmoke, WhaleyTim, WikHead, 3 anonymous edits

Great_Haywood *Source*: http://en.wikipedia.org/w/index.php?title=Great_Haywood *Contributors*: Adambro, AdjustShift, Aidan Croft, Bezapt, Bluesquidharry, Boobalooo, Boomaloo, Carcharoth, CardinalDan, Chris the speller, D6, Dbachmann, De728631, ElectMayor, Fuhghettaboutit, Galoubet, Gjp23, Gonzonoir, Grutness, Headteacher3488, Hk11678, HoraceSlughorn, John of Reading, Madgenberyl, Nono64, Oosoom, Pigsonthewing, Regan123, Rondador, S2SAL, Sally-Ann Sinclair, Sjorford, Spiffy sperry, Tide rolls, WhaleyTim, Zundark, 14 anonymous edits

Little_Haywood *Source*: http://en.wikipedia.org/w/index.php?title=Little_Haywood *Contributors*: Arb, BigrTex, Bobblewik, Boobalooo, Charles Matthews, Colm O'Brien, David Biddulph, Edward, ElectMayor, HeartofaDog, HoraceSlughorn, I-10, Infrogmation, Iridescent, Jza84, NawlinWiki, Outriggr, Polisher of Cobwebs, RHaworth, Regan123, Rekstout, Richard asr, Slugboy, Sophus Bie, Steven J. Anderson, That Guy, From That Show!, TheDarkLordVoldemort, Travelbird, WOSlinker, Winterwater, 19 anonymous edits

National_Trust_for_Places_of_Historic_Interest_or_Natural_Beauty *Source*: http://en.wikipedia.org/w/index.php?title=National_Trust_for_Places_of_Historic_Interest_or_Natural_Beauty *Contributors*: 159753, A.K.A.47, AHMECT, Achilver, Alex Middleton, Alxeedo, Amy Crescenzo, Anetode, Aremith, BD2412, Bankhallbretherton, Bazza 7, Bejnar, Bhoeble, Billinghurst, BjKa, BlueDevil, Bobblewik, BozMo, Brookie, Cactus.man, CalJW, Camboxer, Cameronweaver, Ceyockey, CharlesC, Charlottehenderson, Choalbaton, Chunky Rice, Cnyborg, ConnorJack, Craigy144, Craitman17, D22, Darac, DavidLevinson, Delldot, DinosaursLoveExistence, Dmvward, Dogears, EphraimL, Gabbe, Gaius Cornelius, GearedBull, GrahamBould, Grammarian, Grstain, Hanbrook, Herbythyme, Heron, Hertzsprung, Honbicot, Howcheng, Hu12, Imc, Jamesontai, Jaraalbe, Jdavies, Jdforrester, Jimothytrotter, Jobleblue, John of Reading, Johnbod, JonC, Jooler, Jpbowen, Karlwk, Kbthompson, Kicior99, Kierant, Kwenchin, LPEllis, Lilac Soul, Linuxlad, LukeHoC, MKoltnow, Mais oui!, Mark J, Mark.purcell, Martinevans123, Matthewcl375, Mattis, Mauls, Mgp28, Mhiji, Mhockey, MilborneOne, Miles Pateman, Miyagawa, Montacutemagazine, Morwen, Mpvide65, Mr Taz, Mtaylor848, Nancy, Nationaltrustmag, Ndwood, Nevilley, Now3d, Nskl4, Numbo3, Oliver Chettle, OllieFury, Patrickjoel, Paulmallon, Pcpcpc, Peterkingiron, Pfold, Pigsonthewing, Postlebury, R0pe-196, RJP, Ranger Steve, Regan123, Renata, Renosecond, Rich Farmbrough, Ross Arnold, S19991002, Saga City, Sardanaphalus, SemperBlotto, Shattered, Sidders2007, Skarebo, Slon02, Srikarkashyap, Stj6, Struway2, Superm401, TaerkastUA, Tarquin, Tassedethe, The Thing That Should Not Be, Thegn, Thruston, Tibur, Tommader, Traditional1966, Trident13, Tyrenius, Tyrol5, Ultraexactzz, Wcl2, WikiLaurent, Wikiklrsc, Wimstead, Wknight94, Wprlh, Zscout370, 139 anonymous edits

Earl_of_Lichfield *Source*: http://en.wikipedia.org/w/index.php?title=Earl_of_Lichfield *Contributors*: Aatomic1, Alci12, Anglius, Berks105, Craigy144, Cwitty, Cyan22, DLJessup, Henning M, Histmag, Huw Nathan, Jameswilson, John of Reading, Kittybrewster, Lord Emsworth, Loren Rosen, MarmadukePercy, Mintguy, OwenBlacker, PKM, Phoe, Pilch62, Proteus, Saga City, Scu98rkr, Scwlong, Sitush, Sjorford, Stephen Turner, Triquetra, Tryde, Villafanuk, 19 anonymous edits

Wolseley_Baronets *Source*: http://en.wikipedia.org/w/index.php?title=Wolseley_Baronets *Contributors*: Avicennasis, Ordyg, Phoe, Rich Farmbrough, Tim!, Tryde

English_Benedictine_Congregation *Source*: http://en.wikipedia.org/w/index.php?title=English_Benedictine_Congregation *Contributors*: Briaboru, Carlaude, Charlesdrakew, Daniel the Monk, Greenshed, Handsaw, HeartofaDog, Jaraalbe, Johnpacklambert, Klemen Kocjancic, Lanspergius, Marcusscotus1, Meissen, Mjinkm, Staffelde, The Sage of Stamford, Timtrent, Vivenot, WordyGirl90, 12 anonymous edits

Colwich_Abbey *Source*: http://en.wikipedia.org/w/index.php?title=Colwich_Abbey *Contributors*: Colonies Chris, Deor, Handsaw, Hmains, Lysosome, Mjinkm, Oh Snap, The Sage of Stamford

Shugborough_Hall *Source*: http://en.wikipedia.org/w/index.php?title=Shugborough_Hall *Contributors*: Alivealiveoh, Andy G, Bhoeble, Brinkley32, Bs0u10e01, CambridgeBayWeather, Cameron, Casperonline, Cyan22, Dougweller, EchetusXe, Fkanarya, Gaius Cornelius, Galoubet, Gareth E Kegg, Grstain, Hailey C. Shannon, Hanbrook, J S Ayer, Jetcatflyer, Jllm06, Jonnyboy5, Kcegginton, Keith Edkins, Kk6t, Kurpfalzbilder.de, Lugnuts, Madgenberyl, MarmadukePercy, Matt.whitby, Mav, Mboverload, Megamacht, Mhockey, Misrule13, Moonraker, MortimerCat, Naggie34, Nckidd, Neddyseagoon, Oliver Chettle, Olivier, Oosoom, Ordyg, Pcpcpc, Penrithguy, Plucas58, Regan123, Richardb43, Rls, Saga City, Salming2, Ser Amantio di Nicolao, Sjorford, Smallweed, Son of Paddy's Ego, SpiderJon, Spugmeister, Steinsky, The Cybermacht, Tim!, Villafanuk, Wereon, Wikipediatrix, 33 anonymous edits

Moreton,_Staffordshire *Source*: http://en.wikipedia.org/w/index.php?title=Moreton%2C_Staffordshire *Contributors*: Bnustudent, CanOfWorms, Gadren, Kbthompson, Major monty, Nepahwin, Regan123, Saga City, SarekOfVulcan, Staffelde, 2 anonymous edits

A51_road *Source*: http://en.wikipedia.org/w/index.php?title=A51_road *Contributors*: Blackadderajr, Cacolantern, Deor, Jaraalbe, Jeni, Kbjohnso, Ken Gallager, Mahahahaneapneap, MapsMan, Mauls, Mixsynth, Morwen, Oosoom, Regan123, Severo, Sjorford, TimMassey, Ukeu, WOSlinker, 6 anonymous edits

Southern_Netherlands *Source*: http://en.wikipedia.org/w/index.php?title=Southern_Netherlands *Contributors*: Alansohn, Andrew K. Zimmerman, Angusmclellan, Attilios, Auntof6, Aviad2001, Beetstra, BerndGehrmann, BoH, Booksworm, BrianDaubach10, CSvBibra, Colonies Chris, CommonsDelinker, Cosialscastells, Crowsnest, DVdm, Daisydaisy, Devinlee, Dewritech, Dirk math, Ereunetes, Estlandia, Fabartus, Fefour, Filiep, Fitzwilliam, Fleurstigter, Fuzzbox, Gcm, Ghirlandajo, Gidonb, Goustien, Gryffindor, Gugganij, Hmains, Hutcher, Ilyushka88, Ingolfson, Intangir, J. J. F. Nau, JamesR1701E, Jcegobrain, JhnPlmr, Joelmills, Joeyconnick, John K, Johnpacklambert, Joostik, Joy, Karel Anthonissen, Ketamino, Kusma, Kyng, Lambiam, Leutha, LightPhoenix, Lightmouse, Liist, Lord Cornwallis, M.O.X, MWAK, Malleus Fatuorum, MapsMan, Markussep, Mathiasrex, MatthewVanitas, Member, Mimihitam, Moyogo, Nebuchadnezzar o'neill, Neddyseagoon, Nefertum17, NonZionist, OwenBlacker, Patrick, PaulVIF, Pearle, PhilippWeissenbacher, Phlebas, Pinethicket, Pmaas, Promking, RJP, Rex Germanus, Riana, Rillian, Robofish, RussBlau, SamEV, Shandris, Siafu, Silverhelm, SimonP, SomeHuman, Spot87, Srtxg, Stijn Calle, Str1977, TahitiB, That-Vela-Fella, Tommy2010, Trasamundo, VKokielov, Van helsing, Varlaam, Vnnen, Wechselstrom, Wetman, Whdhshdhshdh, Wikihistorian, 66 anonymous edits

French_Revolution *Source*: http://en.wikipedia.org/w/index.php?title=French_Revolution *Contributors*: 09aandrews, 102, 1122334455, 137.111.13.xxx, 208.19.128.xxx, 2D, 3232330, A. Lafontaine, ABCD, ABF, AP1787, Aaron Schulz, Abductive, Abebhatnagar, Academic Challenger, Acalamari, Adam Bishop, AdamM, Adambro, Adashiel, Adityaput, Admiral Maxtreme, Adrian, Aecis, Ahkond, Ahoerstemeier, AidanPalmer, Akerans, Aksi great, Alapeds, Albertgenii12, Alex Kinloch, Alex S, Alexforcefive, AlexiusHoratius, Alfirin, Alpdpedia, Alphachimp, Alsandro, Altenmann, Amp53194, Ancheta Wis, AndreasJS, Andrei Stroe, Andrew Levine, Andrewpmk, Angusmclellan, Anneman, Anonymous44, Antandrus, Apox, Aprogressivist, Aquaepulse, Arbitrarily0, Archer3, Archer7, Archfalhwyl, Archiesteel, Argo117, Aristotle06, Armatura, ArnoLagrange, ArnoldReinhold, Arrala, Arronax50, Art LaPella, Arwel Parry, Asarelah, Asdfjkl1235, Ashershow1, Astroview120mm, Atlant, Atlantawiki, Aude, Augustes, Avicennasis, Ax0l0tl, Axon, AzaToth, Azerbaijanman, AzureFury, Azza12321, BRUTE, Bachrach44, BadgerBadger, Balsa10, BananaFiend, Banes, Battlesboy, Bbatsell, Bcorr, Bdb484, Bdesham, Belligero, Ben D., Ben-Zin, Bender235, BernardH, Betacommand, Beyond silence, Bhadani, Big Mac Super Mac, Bkwillwm, Blablablob, Blackknight12, Blake324, Blanchardb, Blastwizard, Blightsoot, Blockader, Bloomfield, Bluefire313, Bmicomp, Bob Burkhardt, Bob0044, Bob2561, Bobblewik, Bobbydesi, Bobet, Bobo192, Bodhislutva, Bonadea, Brant.merrell, Breandandalton, Brendan Moody, Brian0918, Brighterorange, Brossow, Bubba hotep, Bubba73, BubbleBabis, Buchanan-Hermit, Buonaparte69, BusterD, Bz2, CQJ, CTZMSC3, Cabalamat, Cactus.man, Cadriel, Cairnsy87, CalebNoble, Cambalachero, Camembert, Camr, Can't sleep, clown will eat me, Can-Dutch, Canadian-Bacon, CanadianCaesar, Canderson7, Canglesea, Canjth, Canterbury Tail, CapitalR, Capricorn42, Carbonite, CardinalDan, Carolina cotton, Carre, CartoonDiablo, CaseyPenk, Casper2k3, Cast, Causa sui, Cbrodersen, Cdc, Celestra, Cenarium, Ceoil, Chairman S., Chameleon, Chaoservices, Chapmlg, Charles Matthews, Charles Nguyen, CharlesGillingham, CharlotteWebb, Charvex, Chatfecter, Chazwozzler, Chederman69, Chinasaur, Chodorkovskiy, Chris O'Riordan, Chris Roy, Chris the speller, Chrislk02, Cinkcool, Ckincaid77, Cleared as filed, ClioFR, ClockworkLunch, ClockworkSoul, Cnollie, Colipon, Color probe, Commander Keane, CommonsDelinker, Computerjoe, Conversion script, Cory Liu, Cote d'Azur, Crcsheedy, Creidieki, Cryptic, Curien, Curly Turkey, Curps, Curtgunz, Cuziyq, Cvaneg, Cyberevil, Cyferx, Cynicism addict, Cyrrk, Czar Brodie, DGJM, DITWIN GRIM, DJ Clayworth, DO'Neil, DTRY, DUBJAY04, DVD R W, Damned Zombie, Dan100, Daniel Quinlan, DanielCD, Danski14, Darolew, Darouet, Darth Panda, Darth ralph wiggum, Dave00327, Dave6, David Joffe, David.Mestel, David.Monniaux, DavidLevinson, DavidWBrooks, Davidalex1, Davodd, Dawes, Dcandeto, Dccarroll, Dcflyer, Deadcorpse, Delldot, Delta Spartan, Dendodge, Denelson83, DennyColt, Dessert fox, Destynereyes, Dfrg.msc, Dgies, Diaa abdelmoneim, Dianaruttman1, Diberri, Didactohedron, DiePerfekteWelle, Din jävla Bög, Dinomite, Discospinster, Diza, Djordjes, Dlohcierekim's sock, Dmlandfair, Dmoss, Doc Tropics, Doc glasgow, Doug Bell, Dr Benway, DrSoucinator, Dravecky, DropShadow, DrunkenSmurf, Dspradau, Dtolman, DuncanHill, Durova, Dustinw88, Dv82matt, Dvptl, Dynasteria, Dyslexik, ESkog, EamonnPKeane, Easy cat, Easyer, Eclecticology, Edison, Edivorce, Edward, Edwy, Eejey, Eilthireach, Ekotkie, ElTyrant, Eldamorie, Eldredo, Elvenscout742, Emre D., Enmerkar, Enviroboy, EoGuy, Epbr123, Eric Sowin, Eric-Wester, Ericamick, Ericd, Erzengel, Eshmart2006, Esoltas, Esperantix, Eurocopter, Evercat, Everyking, Evil Monkey, Explicit, Ezhiki, FF2010, Fab, Fabartus, Factchecker atyourservice, Facts707, Falconclaw5000, FallingRain123, FanG, Fang Aili, Fastfission, Fastman99, Fattyjwoods, Felgs01, FelisLeo, Fernando S. Aldado, Fetofs, FeydHuxtable, Fhjgf, FiP, Firetrap9254, Fisherj.50, Fjmustak, Flannel, Flauto Dolce, FlavrSavr, Flewis, Foraminifera, Formulax, Fram, Francis Burdett, Frangibility, Frania Wisniewska, Frankenpuppy, Freakofnurture, Frederick12, Fredrik, Freemarket, French19, FreplySpang, Frip1000, Fruit666, FullMetal Falcon, Funandtrvl, Func, Fuzheado, Fuzzie, Fvw, Fyver528, G.A.S, GB fan, GHe, GTBacchus, GWatson, Gaara144, Gabeross, Gadfium, Ganems, Gary King, Geneb1955, General Wesc, Geni, Gensanders, Gerber90, Geremia, GhostPirate, Giant dong, Gidonb, Gilliam, Ginkgo100, Gioto, Glen, Gloriamarie, GlosterBoy, Gnaughton, Gomm, Good Olfactory, Gpohara, GraemeL, Graham87, Gramscis cousin, Granpuff, Grayme, Greece666, GreenRoot, Greensburger, Grenavitar, Grika, Grstain, Grules, Gryffindor, Gtdp, Gurch, Guy M, Gwernol, HJ32, HStrout, Hadal, HappyCamper, Harjk, Hashem sfarim, Haukurth, Haverpopper, Havok, Hayabusa future, Headbomb, Heezy, HeikoEvermann, Heimstern, Helli213, Hello32020, HenryPhilipPhd, Henrygb, Hersfold, HexaChord, Hiddekel, Highland14, Hillock65, Historymike, Hockeyman001, Hoopscity, Horesepatjo, Hotlorp, Hro'nyewachu, Hu, HueSatLum, Hun0001, Huon, Husond, Hut 8.5, Huzarus, Hypnoticmonkey, I already forgot, I2ain2t, IRP, IW.HG, IZAK, Iamasuperdragon, Icecold7, Ididit08, Imjustmatthew, Imnotminkus, Impaciente, IndyLawSteve, Infrogmation, Ingolfson, InoShiro, Insanity Incarnate, Inter, InverseHypercube, Iridescent, Irish Plusle, Irish hero, Ishas, Isis, Istvan, Ixfd64, Izehar, Izzymeckler, J.delanoy, J04n, JCarriker, JForget, JHMM13, JJ Georges, JLaTondre, JSKWOOD, JSN2849, JaGa, Jacek Kendysz, Jack Bornholm, Jacktate, Jaeger5432, Jajami, Jan eissfeldt, Janejellyroll, Janhenriegon, Jason.stover, Jasper Deng, Jawed, Jax184, Jayden54, Jburt1, Jeff G.,

Jeffklib, Jeffreymcmanus, Jennifer c martin, JeremyA, Jessicanr, Jezstar, Jiddisch, JinJian, Jlao04, Jmabel, Jmu2108, JoanneB, Joconnor, Joecronin, Jogloran, Johann Wolfgang, John Chamberlain, John K, John Reid, Johnm4, Johntan007, Jojit fb, JonDePlume, Jonathan.s.kt, JorgeGG, Joseph Solis in Australia, Joshsteer, Joshua A Moore, Joshua Boniface, Jossi, Jottinger, Joy, Joyous!, Jpgordon, Jplarkin, Jtdirl, Jtkiefer, Juliancolton, Julius.kusuma, Jvbishop, Jwruffin, KMPalma, Kafka Liz, Kairos, Kaldari, Kaloyan5, Kanags, Kaori, KaragouniS, Karl Dickman, Katefan0, KathrynLybarger, KatinaW, Kbdank71, Kbh3rd, Kcavness, Kcordina, Keegan, Kelly Martin, Keron Cyst, Kerotan, Ketchup krew, Kevin B12, KevinKung, Kevinfromhk, Kfodderst, Khalid Mahmood, Khalid hassani, Khenbish, Khfan93, Kieran4, Kilo-Lima, Kilrogdeadeye, Kingturtle, Kinneyboy90, Kizor, KnightRider, KnowledgeOfSelf, Knutux, Kondspi, Konstable, Korg, Kowey, Koyaanis Qatsi, Kozuch, Kpalion, KrakatoaKatie, Kralahome, Krich, Kross, Ksenon, Kubigula, Kungfuadam, Kuru, Kurykh, Kzollman, LAX, Labine50, Lacrimosus, Lady of Versailles, Lakers, Larry2006, Laurascudder, Law, Lazulilasher, Leafyplant, LeaveSleaves, Lectonar, Ledenierhomme, Leinad-Z, Lemmey, LeoDV, Leszek Jańczuk, Letsdancebabe, LifeStar, Liftarn, Lightdarkness, Lightmouse, Lilaznpookid, Linkminer, Linnell, Lionel Elie Mamane, Litefantastic, LittleOldMe, Lkmorlan, Llort, Llywrch, Lockesdonkey, Logan, Lonely loner, Longshot.222, Looxix, Lord Cornwallis, Loretta Fairfechen, Lorzu, Lottamiata, Louiegarcia 3@hotmail.com, Lucent, Luigi30, Lumaga, Luna Santin, Lupo, Lycurgus, M.Gray, MC10, MER-C, MONGO, MWatson15, Mabahandula, Macattack5436, Madman2001, Maelnuneb, Magioladitis, Magister Mathematicae, Malchemist, Malcolm, Mallocks, Mamalujo, Man vyi, Manjunathsinge, Mark Heiden, MarsRover, Martial75, Massimamanno, Master Jay, Master2841, Matia.gr, Mato, Matthead, Matthew Yeager, Matusz, Mav, Max Thayer, Maxamegalon2000, Maxis ftw, Mazin07, Mbc362, Mcohan, Mdukas, Mebden, Mediareport, Meelar, Meeso, Melaen, Melchoir, Metageek, Metropolitan, Mets501, Michael Hardy, Michaelmas1957, Middaythought, MidgleyDJ, Mike Rosoft, Mikko Paananen, Misza13, MjMenuet111, Mjpieters, Mjs110, Mkoyle, Modernist, Modest Genius, Moe Epsilon, Moeron, Moink, Molly-in-md, Moneyball026, Moreschi, Morwen, Motorizer, Mpeisenbr, Mr Adequate, Mr Stephen, Mr. Know-It-All, Mr. Lefty, MrFish, Mxn, Mygerardromance, Mzzl, N2e, NYArtsnWords, Naddy, Naive cynic, Nakon, Nat, Nathan.tang, Naufana, Nauticashades, NawlinWiki, Nbisrainbow, Ncmvocalist, Neddyseagoon, NeilN, Neoncow, Nesbitt, Neuroscientist, Nick Number, NightFalcon90909, Nilfanion, Nishkid64, Nk, No Guru, NoIdeaNick, NoSeptember, Nolege, Notchcode, Notepad, Nova12gauge, Novacatz, NubKnacker, Numbo3, Nyh, O0pyromancer0o, Obamamaniac, Obli, Oceanblueeyes, Oda Mari, Ohconfucius, Okapi, Oleg Alexandrov, Olivier, Omicronpersei8, Orangutan, OrbitOne, OreL.D, Ospalh, Otolemur crassicaudatus, OttawaAC, Ouro, Outriggr, OwenX, P0k0mYd0nGk0, PAntoni, PHDrillSergeant, Paelius, Pagliuco, Paines, Pakaran, Palthrow, Paluki, Parthian Scribe, Pat Payne, Patrick, Patstuart, Paul August, Paulscf, Pavel Vozenilek, Pb30, Pcbene, Pedant17, Pentium4, Peregrine981, Persian Poet Gal, Peruvianllama, PeterBFZ, Pethr, Petter Bøckman, Pgan002, Pgk, Pharaoh of the Wizards, Phoebus, Phoenix Hacker, Phoenix2, Photographerguy, Photonikonman, Pi3832, Piano non troppo, Piccolo Modificatore Laborioso, Pilotguy, Pippa12, PoccilScript, Poetaris, Poodaboo, Poor Yorick, Pootown, Postdlf, Prolog, PrometheusX303, Proteus, Pseudo-Richard, Psy guy, Puchiko, PureRED, Purplethinker, Purpleturple, Pwanyonyi, Quantum Mage, Quendus, R'n'B, R. fiend, R.small, RJaguar3, RUorange87, Rama's Arrow, Ramiki, Ramsis II, Ranmoth, Ranveig, Raven in Orbit, Raymond, Rdsmith4, Reaper X, Red Dalek, Red Director, Reddi, Redonkulous, Reedy, RegRCN, Remember, Renaissanceboy, Renamed user 1752, Renski, RepublicanJacobite, Retired username, Rettetast, Rex Germanus, RexNL, RexxS, Rholton, Riana, Ricelovva, Rich Farmbrough, RichiH, Richwales, Rickyrab, Riddler, Rivkid007, Rjensen, Rjwilmsi, Rklawton, Rmhermen, Rmt2m, RobertG, Robertvan1, Robin2550, Robomaeyhem, RodC, Roentgenium111, Rogswikipage, Romanhouse, Romanm, Romarin, Ronabop, Ronnyim12345, Rory096, RoyBoy, Royboycrashfan, Rs09985, Rudjek, Ruhrjung, Rune.welsh, Rursus, RussBlau, Ruy Lopez, Ryanbomber, Ryanw315, Ryomaandres, Ryulong, SFGiants, SHIMONSHA, Saddhiyama, SallyBoseman, SamJ7792, Samuelsen, Sanfranman59, Sango123, Sanmartin, Sannse, Science4sail, Scott5834, Scwlong, SeanSchricker, Searley, Seherr, Seidenstud, Sekhar 06, Seryo93, Shadowblade34, Shadowjams, Shanel, Shanes, Shearonink, Shirulashem, Shizane, Shizhao, Shoessss, Sickjohnny, Sifaka, Silent reverie86, Silverhorse, SimonP, SineWave, Sionus, Sir Nicholas de Mimsy-Porpington, Sjakkalle, Sjc, Skyring, Slgrandson, Slowking Man, Slrubenstein, Smack, Smmurphy, Snkla2, Snowdog, Snoyes, SoLando, Soccerboy88, Solitude, Soman, Someguy1221, Someone else, SpNeo, Spangineer, Spellcast, Spencer F., Splash, SpookyMulder, Spoonkymonkey, Squeezeweasel, Squell, Squiddy, Squideshi, Srsjones, St.daniel, Stars4change, StealthEXE, Stephenb, Stephfo, Steve Farrell, SteveStrummer, Stevietheman, Stjust11, Stmccoy, Stoo1981, StrangeAttractor, SuperJumbo, SuperMoonMan, Swartik, Swedish fusilier, SweetNeo85, Synergy, SyntaxError55, Syrthiss, Sysys, T-bonham, T-man33, TBadger, TS1, TWallyT, Tabw369, Taco325i, Tadartwb, Tagishsimon, Talon Artaine, Tangotango, Taraalexa3, Taral, Tariqabjotu, Tarquin, Tarret, Tawker, Tazmaniacs, Tbone55, Ted Ted, Teeheeee, Tellkel, Template namespace initialisation script, TerritorialWaters, TexasAndroid, Th1rt3en, The Anome, The Deviant, The Enlightened, The Four Deuces, The Fwanksta, The Rabbit42, The Rambling Man, The2man, TheDarkArchon, TheKMan, Thecheesykid, Thedjatclubrock, Thefourdotelipsis, Thehalfone, Thelawnmower69, Themightyquill, Theon, Thesis4Eva, Thingg, Thomas Larsen, Thomcbh, ThreeSpears, Thue, Thunderboltz, TigerShark, TimMartin, Timbouctou, Titoxd, Tlogmer, Tobby72, Todeswalzer, Tohd8BohaithuGh1, Tom fletcher, Tom harrison, TomPhil, Tomdatom, Tomlillis, Tommasar, Tommibg, Tony Fox, Tony Sidaway, Tony1, Tonyob, Top.Squark, Tothebarricades.tk, Tpbradbury, TreasuryTag, Tsemii, Tswsl1989, Tubestick, TulsaTown, Tumblingsky, Tyler, TylerC, UberCryxic, Ugur Basak, UkPaolo, Ultimaga, Ultimate Destiny, Ultracoolxxx, Ultron, Unclerodya, UninvitedCompany, Unlockitall1, Unobanana, Upi, Urhixidur, Urzadek, Useless Fodder, User2004, VBGFscJUn3, Vancouver Outlaw, Vanessa Robinson, Vary, Vasile, Veinor, Veronicamars13, Vincent Lextrait, Viridian, Vis-a-visconti, Viscount, Viskonsas, Voceditenore, VolatileChemical, Voofie, Vrael**, Vsmith, WAvegetarian, WGee, WHEELER, WJBscribe, Wafulz, Waggers, Wahabijaz, Walton One, Wapcaplet, Warofdreams, Water falls are pertty, Wavelength, Wayward, Wechselstrom, Wedineinheck, Weena Eloi, Wenli, Wenteng, Wes!, West Brom 4ever, WestFullMoon, Wetman, Weyes, Whippasnipa, WhisperToMe, Whuzap, Wikedguy, Wiki alf, Wikieditor06, Wikikris, WikipedianMarlith, Wikiscrewup, Will Beback Auto, Willworkforicecream, Wilt, Wimt, WinstonSmith, Wizzy, Wmahan, Wolfman, Woohookitty, Worik, WowLabia, Wowomg.ca, Writtenright, Www.wikinerds.org, Xandar, Xaosflux, Xchbla423, Xcube64, Xdenizen, Xerocs, Xezbeth, Xiahou, Xiner, Xsahilx, Yamamoto Ichiro, YankeeDoodle14, YellowMonkey, Yjl, Ynhockey, YnnusOiramo, Yohan euan o4, Yuckfoo, Yug, Zargulon, Zastard, Zdravko mk, Zessa, Zetud, Zoicon5, Zondor, ^Jedi^Gen.krazy karateka, 3220 anonymous edits

Image Sources, Licenses and Contributors

Image:Haywood Junction.jpg *Source*: http://en.wikipedia.org/w/index.php?title=File:Haywood_Junction.jpg *License*: unknown *Contributors*: Gjp23 at en.wikipedia

file:Staffordshire UK location map.svg *Source*: http://en.wikipedia.org/w/index.php?title=File:Staffordshire_UK_location_map.svg *License*: unknown *Contributors*: User:Nilfanion

File:Red pog.svg *Source*: http://en.wikipedia.org/w/index.php?title=File:Red_pog.svg *License*: unknown *Contributors*: Anomie

Image:National Trust logo.png *Source*: http://en.wikipedia.org/w/index.php?title=File:National_Trust_logo.png *License*: unknown *Contributors*: Ndwood, Sfan00 IMG, Tyrol5

File:Alfriston Clergy House oak leaf carving 2007.JPG *Source*: http://en.wikipedia.org/w/index.php?title=File:Alfriston_Clergy_House_oak_leaf_carving_2007.JPG *License*: unknown *Contributors*: GrahamBould, Kurpfalzbilder.de

File:Wicken Lode1.JPG *Source*: http://en.wikipedia.org/w/index.php?title=File:Wicken_Lode1.JPG *License*: unknown *Contributors*: User:Naturenet

File:National Trust at Gordale.JPG *Source*: http://en.wikipedia.org/w/index.php?title=File:National_Trust_at_Gordale.JPG *License*: unknown *Contributors*: User:Immanuel Giel

File:UK-National-Trust-signpost.jpg *Source*: http://en.wikipedia.org/w/index.php?title=File:UK-National-Trust-signpost.jpg *License*: unknown *Contributors*: Toby Thurston

File:Worm's Head (Rhossili).jpg *Source*: http://en.wikipedia.org/w/index.php?title=File:Worm's_Head_(Rhossili).jpg *License*: unknown *Contributors*: Original uploader was CharlesC at en.wikipedia

File:Stourhead Garden View from Above.JPG *Source*: http://en.wikipedia.org/w/index.php?title=File:Stourhead_Garden_View_from_Above.JPG *License*: unknown *Contributors*: Original uploader was Daderot at en.wikipedia

File:WaddesdonManor.JPG *Source*: http://en.wikipedia.org/w/index.php?title=File:WaddesdonManor.JPG *License*: unknown *Contributors*: Mattlever

Image:DitchleyHousemorris.jpg *Source*: http://en.wikipedia.org/w/index.php?title=File:DitchleyHousemorris.jpg *License*: unknown *Contributors*: Merchbow, Welbeck

Image:Shugborough Hall Jones' Views 1829.jpg *Source*: http://en.wikipedia.org/w/index.php?title=File:Shugborough_Hall_Jones'_Views_1829.jpg *License*: unknown *Contributors*: Bhoeble, Kurpfalzbilder.de, Ratarsed

Image:Shugborough Hall 01.jpg *Source*: http://en.wikipedia.org/w/index.php?title=File:Shugborough_Hall_01.jpg *License*: unknown *Contributors*: Mandy Moore

Image:Anson-Gosse-1750-24.jpg *Source*: http://en.wikipedia.org/w/index.php?title=File:Anson-Gosse-1750-24.jpg *License*: unknown *Contributors*: non identifié

Image:Shugborough inscription.jpg *Source*: http://en.wikipedia.org/w/index.php?title=File:Shugborough_inscription.jpg *License*: unknown *Contributors*: Naggie34, Rauenstein

Image:Shugborough arcadia.jpg *Source*: http://en.wikipedia.org/w/index.php?title=File:Shugborough_arcadia.jpg *License*: unknown *Contributors*: Bohr, Migel Sances Huares, Rauenstein, Wst

File:UK road A51.svg *Source*: http://en.wikipedia.org/w/index.php?title=File:UK_road_A51.svg *License*: unknown *Contributors*: User:Mauls

Image:UK road A56.svg *Source*: http://en.wikipedia.org/w/index.php?title=File:UK_road_A56.svg *License*: unknown *Contributors*: Mauls

Image:UK road A5268.PNG *Source*: http://en.wikipedia.org/w/index.php?title=File:UK_road_A5268.PNG *License*: unknown *Contributors*: Regan123 at en.wikipedia

Image:UK road A41.svg *Source*: http://en.wikipedia.org/w/index.php?title=File:UK_road_A41.svg *License*: unknown *Contributors*: User:Mauls

Image:UK road A55.svg *Source*: http://en.wikipedia.org/w/index.php?title=File:UK_road_A55.svg *License*: unknown *Contributors*: User:Mauls

Image:UK road A54.svg *Source*: http://en.wikipedia.org/w/index.php?title=File:UK_road_A54.svg *License*: unknown *Contributors*: User:Mauls

Image:UK road A49.svg *Source*: http://en.wikipedia.org/w/index.php?title=File:UK_road_A49.svg *License*: unknown *Contributors*: User:Mauls

Image:UK road A534.PNG *Source*: http://en.wikipedia.org/w/index.php?title=File:UK_road_A534.PNG *License*: unknown *Contributors*: Regan123 at en.wikipedia

Image:UK road A500.PNG *Source*: http://en.wikipedia.org/w/index.php?title=File:UK_road_A500.PNG *License*: unknown *Contributors*: Regan123 at en.wikipedia

Image:UK road A525.PNG *Source*: http://en.wikipedia.org/w/index.php?title=File:UK_road_A525.PNG *License*: unknown *Contributors*: Regan123 at en.wikipedia

Image:UK road A53.svg *Source*: http://en.wikipedia.org/w/index.php?title=File:UK_road_A53.svg *License*: unknown *Contributors*: User:Mauls

Image:UK road A34.svg *Source*: http://en.wikipedia.org/w/index.php?title=File:UK_road_A34.svg *License*: unknown *Contributors*: User:Mauls

Image:UK road A518.PNG *Source*: http://en.wikipedia.org/w/index.php?title=File:UK_road_A518.PNG *License*: unknown *Contributors*: Regan123 at en.wikipedia

Image:UK road A513.PNG *Source*: http://en.wikipedia.org/w/index.php?title=File:UK_road_A513.PNG *License*: unknown *Contributors*: Regan123 at en.wikipedia

Image:UK road A469.PNG *Source*: http://en.wikipedia.org/w/index.php?title=File:UK_road_A469.PNG *License*: unknown *Contributors*: Regan123 at en.wikipedia

Image:UK road A5192.PNG *Source*: http://en.wikipedia.org/w/index.php?title=File:UK_road_A5192.PNG *License*: unknown *Contributors*: Regan123 at en.wikipedia

Image:UK road A5.svg *Source*: http://en.wikipedia.org/w/index.php?title=File:UK_road_A5.svg *License*: unknown *Contributors*: User:Mauls

File:Arms of Flanders.svg *Source*: http://en.wikipedia.org/w/index.php?title=File:Arms_of_Flanders.svg *License*: unknown *Contributors*: Adelbrecht, Rinaldum, 1 anonymous edits

File:Armoiries Principauté de Liège.svg *Source*: http://en.wikipedia.org/w/index.php?title=File:Armoiries_Principauté_de_Liège.svg *License*: unknown *Contributors*: User:Odejea

File:Gules a fess argent.svg *Source*: http://en.wikipedia.org/w/index.php?title=File:Gules_a_fess_argent.svg *License*: unknown *Contributors*: user:Odejea

File:Banner of the Holy Roman Emperor with haloes (1400-1806).svg *Source*: http://en.wikipedia.org/w/index.php?title=File:Banner_of_the_Holy_Roman_Emperor_with_haloes_(1400-1806).svg *License*: unknown *Contributors*: User:David Liuzzo, User:N3MO

File:Arms of the king of the Belgians (since 1921).svg *Source*: http://en.wikipedia.org/w/index.php?title=File:Arms_of_the_king_of_the_Belgians_(since_1921).svg *License*: unknown *Contributors*: User:Katepanomegas

File:Arms of the Counts of Luxembourg.svg *Source*: http://en.wikipedia.org/w/index.php?title=File:Arms_of_the_Counts_of_Luxembourg.svg *License*: unknown *Contributors*: User:Adelbrecht

File:Counts of Holland Arms.svg *Source*: http://en.wikipedia.org/w/index.php?title=File:Counts_of_Holland_Arms.svg *License*: unknown *Contributors*: User:Ipankonin

File:Arms of the Duke of Burgundy (1364-1404).svg *Source*: http://en.wikipedia.org/w/index.php?title=File:Arms_of_the_Duke_of_Burgundy_(1364-1404).svg *License*: unknown *Contributors*: User:Heralder

File:Flag of the Low Countries.svg *Source*: http://en.wikipedia.org/w/index.php?title=File:Flag_of_the_Low_Countries.svg *License*: unknown *Contributors*: User:Adelbrecht

File:Prinsenvlag.svg *Source*: http://en.wikipedia.org/w/index.php?title=File:Prinsenvlag.svg *License*: unknown *Contributors*: Artem Karimov, Bender235, Cirt, Golradir, Havang(nl), Homo lupus, JdH, Ludger1961, Miyamaki, Oren neu dag, Rene Bekker, Sceptic, Vincent Steenberg, 4 anonymous edits

File:Flag of Austrian Low Countries.svg *Source*: http://en.wikipedia.org/w/index.php?title=File:Flag_of_Austrian_Low_Countries.svg *License*: unknown *Contributors*: User:Oren neu dag

File:LuikVlag.svg *Source*: http://en.wikipedia.org/w/index.php?title=File:LuikVlag.svg *License*: unknown *Contributors*: User:Westermarck

File:Flag of the Brabantine Revolution.svg *Source*: http://en.wikipedia.org/w/index.php?title=File:Flag_of_the_Brabantine_Revolution.svg *License*: unknown *Contributors*: User:Anno16

File:Flag of France.svg *Source*: http://en.wikipedia.org/w/index.php?title=File:Flag_of_France.svg *License*: unknown *Contributors*: Anomie

File:Flag of the Batavian Republic.svg *Source*: http://en.wikipedia.org/w/index.php?title=File:Flag_of_the_Batavian_Republic.svg *License*: unknown *Contributors*: User:Adelbrecht

Image:Flag of the Netherlands.svg *Source*: http://en.wikipedia.org/w/index.php?title=File:Flag_of_the_Netherlands.svg *License*: unknown *Contributors*: User:Zscout370

File:Flag of the Netherlands.svg *Source*: http://en.wikipedia.org/w/index.php?title=File:Flag_of_the_Netherlands.svg *License*: unknown *Contributors*: User:Zscout370

File:Flag of Belgium.svg *Source*: http://en.wikipedia.org/w/index.php?title=File:Flag_of_Belgium.svg *License*: unknown *Contributors*: User:Dbenbenn

File:Flag of Luxembourg.svg *Source*: http://en.wikipedia.org/w/index.php?title=File:Flag_of_Luxembourg.svg *License*: unknown *Contributors*: User:SKopp

Image:Habsburg Map 1547.jpg *Source*: http://en.wikipedia.org/w/index.php?title=File:Habsburg_Map_1547.jpg *License*: unknown *Contributors*: edited by Sir Adolphus William Ward, G.W. Prothero, Sir Stanley Mordaunt Leathes

Image:Espagnols.PNG *Source*: http://en.wikipedia.org/w/index.php?title=File:Espagnols.PNG *License*: unknown *Contributors*: Bender235, BrightRaven, Cycn, Electionworld, Erik Warmelink, Paul2, Vincent Steenberg

File:Prise de la Bastille.jpg *Source*: http://en.wikipedia.org/w/index.php?title=File:Prise_de_la_Bastille.jpg *License*: unknown *Contributors*: Jean-Pierre Houël (1735-1813)

File:Ludvig XVI av Frankrike porträtterad av AF Callet.jpg *Source*: http://en.wikipedia.org/w/index.php?title=File:Ludvig_XVI_av_Frankrike_porträtterad_av_AF_Callet.jpg *License*: unknown *Contributors*: AnRo0002, BLueFiSH.as, Carolus, Dcoetzee, Ecummenic, Havang(nl), Homo lupus, Kaho Mitsuki, Kuerschner, Kürschner, MARKELLOS, Mathiasrex, Shakko, Thorvaldsson

File:Troisordres.jpg *Source*: http://en.wikipedia.org/w/index.php?title=File:Troisordres.jpg *License*: unknown *Contributors*: Anne97432, AnonMoos, Baronnet, Gryffindor, Jospe, Man vyi, Shakko, Skipjack, W. C. Minor, 2 anonymous edits

File:Estatesgeneral.jpg *Source*: http://en.wikipedia.org/w/index.php?title=File:Estatesgeneral.jpg *License*: unknown *Contributors*: J.M. Moreau Le Jeune.

Image:Le Serment du Jeu de paume.jpg *Source*: http://en.wikipedia.org/w/index.php?title=File:Le_Serment_du_Jeu_de_paume.jpg *License*: unknown *Contributors*: AnRo0002, Anne97432, Bukk, Coyau, Herbythyme, Kirtap, Man vyi, Martin H., Mu, Mutter Erde, Paola Severi Michelangeli, Sammyday, Tablar, 9 anonymous edits

Image:Declaration of Human Rights.jpg *Source*: http://en.wikipedia.org/w/index.php?title=File:Declaration_of_Human_Rights.jpg *License*: unknown *Contributors*: -

File:Women's March on Versailles01.jpg *Source*: http://en.wikipedia.org/w/index.php?title=File:Women's_March_on_Versailles01.jpg *License*: unknown *Contributors*: Anne97432, Interpretix, Kaganer

Image:Decret de l'Assemblée National qui supprime les Ordres Religieux et Religieuses.jpg *Source*: http://en.wikipedia.org/w/index.php?title=File:Decret_de_l'Assemblée_National_qui_supprime_les_Ordres_Religieux_et_Religieuses.jpg *License*: unknown *Contributors*: non identifié / unknown

Image:Federation.jpg *Source*: http://en.wikipedia.org/w/index.php?title=File:Federation.jpg *License*: unknown *Contributors*: Berrucomons, Olivier2, Paris 16, Pmx, Rama, Ultimate Destiny, Vberger

File:Retour Varennes 1791.jpg *Source*: http://en.wikipedia.org/w/index.php?title=File:Retour_Varennes_1791.jpg *License*: unknown *Contributors*: unknown

Image:Jacques Bertaux - Prise du palais des Tuileries - 1793 .jpg *Source*: http://en.wikipedia.org/w/index.php?title=File:Jacques_Bertaux_-_Prise_du_palais_des_Tuileries_-_1793_.jpg *License*: unknown *Contributors*: Anne97432, Bukk, Coyau, Ecummenic, Equendil, Gödeke, 1 anonymous edits

Image:LouisXVIExecutionBig.jpg *Source*: http://en.wikipedia.org/w/index.php?title=File:LouisXVIExecutionBig.jpg *License*: unknown *Contributors*: AnRo0002, Anne97432, Bkwillwm, Bohème, Cohesion, Herbythyme, J JMesserly, Man vyi, Mu, Paris 16, Pmx, Pyrope, Tangopaso, Trycatch, Ultimate Destiny, 9 anonymous edits

Image:Cruikshank - The Radical's Arms.png *Source*: http://en.wikipedia.org/w/index.php?title=File:Cruikshank_-_The_Radical's_Arms.png *License*: unknown *Contributors*: Anne97432, Bohème, Dahn, Ecummenic, Erri4a, Frank Schulenburg, Wolfmann

File:Jacques-Louis David - Marie Antoinette on the Way to the Guillotine.jpg *Source*: http://en.wikipedia.org/w/index.php?title=File:Jacques-Louis_David_-_Marie_Antoinette_on_the_Way_to_the_Guillotine.jpg *License*: unknown *Contributors*: AnRo0002, Anne97432, Gryffindor, Kirtap, Melesse, 3 anonymous edits

File:BatailleduMans1793.jpg *Source*: http://en.wikipedia.org/w/index.php?title=File:BatailleduMans1793.jpg *License*: unknown *Contributors*: AnRo0002, Anne97432, Barbe-Noire, Bukk, Diomede, Khaerr, Mathiasrex, Mu, Olivier2, Pymouss

File:Fête de l'Etre suprême 1.jpg *Source*: http://en.wikipedia.org/w/index.php?title=File:Fête_de_l'Etre_suprême_1.jpg *License*: unknown *Contributors*: Alexandrin, AnRo0002, Anne97432, Civa, Ellin Beltz, Interpretix, Kilom691, Olivier2, Paris 16, Tablar, Ultimate Destiny, 1 anonymous edits

File:Execution robespierre, saint just....jpg *Source*: http://en.wikipedia.org/w/index.php?title=File:Execution_robespierre,_saint_just....jpg *License*: unknown *Contributors*: Ambre Troizat, AnRo0002, Anne97432, Bohème, DIREKTOR, David Kernow, J JMesserly, Jafeluv, Kalki, Krinkle, Maksim, Man vyi, Olivier2, Pmx, Svencb, Tangopaso, Trycatch, Ultimate Destiny, Wst, 3 anonymous edits

Image:Bouchot - Le general Bonaparte au Conseil des Cinq-Cents.jpg *Source*: http://en.wikipedia.org/w/index.php?title=File:Bouchot_-_Le_general_Bonaparte_au_Conseil_des_Cinq-Cents.jpg *License*: unknown *Contributors*: -Strogoff-, ANGELUS, AnRo0002, Anne97432, Bukk, Coyau, Evrik, Flominator, Frank Schulenburg, Ketamino, Kilom691, Luigi Chiesa, Man vyi, Mattes, Mu, Olivier2, Parisette, Place Clichy, TheJH, Xhienne, 3 anonymous edits

Image:Sans-culotte.jpg *Source*: http://en.wikipedia.org/w/index.php?title=File:Sans-culotte.jpg *License*: unknown *Contributors*: Aurevilly, FA2010, James086, Javierme, Kilom691, Mu, Olivier2, TwoWings

File:Marie-Olympe-de-Gouges.jpg *Source*: http://en.wikipedia.org/w/index.php?title=File:Marie-Olympe-de-Gouges.jpg *License*: unknown *Contributors*: Awadewit, Badzil, Beria, Ecummenic, Jaucourt, 1 anonymous edits

Printed by Books on Demand GmbH, Norderstedt / Germany